# The River of Life

Ecosystem Science and Applications

**EDITORS**

Jiquan Chen

Heidi Asbjornsen

Kristiina A. Vogt

# The River of Life

## Sustainable Practices of Native Americans and Indigenous Peoples

Edited by Michael E. Marchand, Kristiina A. Vogt, Asep S. Suntana, Rodney Cawston, John C. Gordon, Mia Siscawati, Daniel J. Vogt, John D. Tovey, Ragnhildur Sigurdardottir, Patricia A. Roads

With contributions by Wendell George, John McCoy, Melody Starya Mobley, Jonathan Tallman, Ryan Rosendal, Cheryl Grunlose

Michigan State University Press | *East Lansing*

∞ The paper used in this publication meets the minimum requirements of ANSI/NISO Z39.48-1992 (R 1997) (Permanence of Paper).

Michigan State University Press
East Lansing, Michigan 48823-5245

Printed and bound in the United States of America.

25 24 23 22 21 20 19 18 17 16    1 2 3 4 5 6 7 8 9 10

LIBRARY OF CONGRESS CATALOGING-IN-PUBLICATION DATA
Marchand, Michael E.
The river of life : sustainable practices of Native Americans and indigenous peoples / Michael E. Marchand, Kristiina A. Vogt, Asep S. Suntana, Rodney Cawston, John C. Gordon, Mia Siscawati, Daniel J. Vogt, John D. Tovey, Ragnhildur Sigurdardottir, Patricia A. Roads ; with contributions by Wendell George, John McCoy, Melody Starya Mobley, Jonathan Tallman, Ryan Rosendal, Cheryl Grunlose.
Includes bibliographical references (pages 257–266) and index.
ISBN 978-1-61186-222-5 (pbk. : alk. paper)
1. Indians of North America—Science. 2. Indian philosophy North America. 3. Indigenous peoples—Ecology—North America. 4. Traditional ecological knowledge—North America. 5. Sustainable living—North America. 6. Sustainable development—North America. I. Title.
E98.S43 M37 2014
970.004'97—dc23
2013036778

Cover design by Shaun Allshouse, www.shaunallshouse.com
Cover image provided by Cheryl A. Grunlose

green press INITIATIVE Michigan State University Press is a member of the Green Press Initiative and is committed to developing and encouraging ecologically responsible publishing practices. For more information about the Green Press Initiative and the use of recycled paper in book publishing, please visit www.greenpressinitiative.org.

Visit Michigan State University Press at *www.msupress.org*

# Preface

Sustainability is a concept that really registered on the consciousness of global societies in the late 1980s [1]. This word defines the need for society to live within the constraints of the land's capacity to deliver the fossil and natural resources that society consumes. Since volumes of materials have been written on "sustainability" by a multitude of authors, and all with different views, one can ask the logical question: "Why has a robust road map to implement 'sustainable practices' not evolved from all this knowledge?" Do we need another book on this topic? We argue that the answer to this question is a resounding **Yes, Yes, Yes**. But such a book should not repeat and summarize what has already been so eloquently authored.

Our belief is that sustainability practices need to include the diversity of nature practices and cultural norms that are passed down inter-generationally by native peoples. When global communities industrialize, they lose their cultural practices and inter-generational knowledge that retains their connections to nature. We contend for industrializing countries to be "real" sustainable practitioners, they need to re-establish their cultural links to nature. They should not completely transform themselves to look, speak and write like a "global citizen" who is computer and technology savvy but has no local culture linked to nature. An industrial country citizen is a person who has joined the **melting pot** to become a generic citizen of the world, e.g., someone who dresses the same way, eats the same foods, lives in the same style housing, owns a cell phone with the latest technology, etc. A global citizen suffers from a "nature deficit" syndrome, lacks knowledge of the impacts of their land-use activities on nature and nature is either desired for its economic benefits or for its ecosystem services.

The western world citizen is decoupled from nature and uses their engineering ingenuity to harvest resources from nature during times of plenty and scarcity. Technology allows western societies to push nature beyond its thresholds and for people to live comfortably under climate-controlled conditions. Scarcity is less of a problem for society to deal with. Society is less worried about making competitive and difficult choices since natural resources or their substitutes are abundant and do not limit economic development. Inter-generational knowledge is not needed since human engineering can always design a machine or produce a chemical to make life easier for people.

The western world life is becoming more complicated and difficult to live today. Some of the technologies have social and environmental impacts that society is unable to avoid. These unintended environmental and social consequences were not on societies radar screen when the wonders of new technology were initially introduced. Now the industrialized citizen is vulnerable to land-use and climate change introduced by some of these technologies. Today, society cannot move elsewhere to avoid the impacts of these technologies. Furthermore, taking resources from someone else during periods of scarcity does not work anymore.

Today, resource scarcity and environmental degradation is on the western worlds radar screen. Now an industrialized world citizen has to collect and consume resources during both nature's boom and bust cycles. A toolkit dependent only on technology solutions and economic models does not work when resource abundance is cyclic. Fortunately for the western world, native people are adapted to nature's boom and bust cycles. Native people also use a holistic approach to simultaneously make environmental, social and economic decisions. They adapt to and respect nature's cycles of resource abundance and scarcity (see section 10). They have much to teach the western world.

Humanizing sustainable practices are not going to be easy for the western world citizen. The need to "ensure a balance between economic development, social development and environmental protections as interdependent and mutually reinforcing pillars of sustainable development" requires each person to become a holistic thinker and planner [2]. Even though the debates and discussions are occurring on sustainable practices, the western world is still at the conceptual stage to humanize sustainability practices.

The western world has mostly developed a multitude of robust economic models to make economic trade-offs [1]. But the story cannot end here. Economic assessments appear to provide clear options for decision-makers but inevitably someone gains from selling resources while someone else loses when these transactions occur. This is not an equitable use of resources and, when economics dictate decision-making, nature is generally ignored in the process. It is not uncommon for decision-making to be riddled with conflicts and debates. This situation increases the possibility that valid opposing viewpoints are not heard. Instead, decisions are made using selective data sources, i.e., decision-making by special interest groups.

Special interest groups are effective at using economic data in quantitative models to support their views. For example, costs and benefits are routinely calculated as part of economic models. It appears that all the options are sufficiently vetted by the experts conducting these assessments because mathematics makes economic tools credible. This does not address the problem that mathematical decision models are limited to comparing three different variables at one time [3]. This means that someone has to decide which variables are included in these analyses and therefore what data are selected for input into these models.

When economic models are the dominant tool used to assess sustainable practices, it is not uncommon for people living in developing countries (particularly indigenous peoples) to receive fewer benefits. This situation can occur even when these people are the suppliers of resources to the industrialized countries. The logical assumption would be that people who own the resources would receive more benefits than those societies that distribute or consume resources. This rarely happens since the western world economic models are structured to benefit industrialized- and technology-based societies.

The International Forum on Globalization (IFG) claims that globalization is not merely a question of marginalizing indigenous peoples but is a multi-pronged attack on the very foundation of their existence and livelihoods [4]. Indigenous people mainly bear the brunt of the costs (low compensation and degrading environments) when nature is harvested for economic gain. The environmental and social repercussions are felt more strongly by the resource-supplier countries. These countries retain fewer viable options when their lands are over-exploited for resources since they still depend upon these same lands for their sustenance and survival. Resource scarcity becomes even a greater problem if the health of the lands decreases following over-exploitation.

Scarcity of resources is not a new issue for society to deal with. It was debated more than one hundred years ago. Then western societies predicted resource scarcity was going to cause the collapse of the global population and economies. However, this gloom and doom did not happen. Human ingenuity and technological developments increased food production yields from the same piece of land. Technological developments allowed societies to increase resource supplies beyond the land's capacity to grow it naturally; e.g., fossil fuel-based fertilizers and herbicides allowed crop yields to increase beyond the natural productive capacity of each land and the limits of each soil and climate. Western societies also used resources more efficiently. This meant that a larger population could be fed and the predicted gloom and doom cycles became a thing of the past.

Scarcity of resources is not just a problem of the past. It will reoccur to impact societies when poor trade-off decisions are made. In the past, technological advancements did not consider the societal and environmental externalities of implementing a new technology. The technology is not bad but how it is used can become bad for the environment and society. A common source of negative externalities is over-exploitation of nature's resource capital or enhancing a land's productive capacity using technology that decreases the land's resilience. There may be a point where technology is less able to compensate for the loss of lands resilience. We are not saying that we need to avoid technology but it needs to be implemented in a manner that does not reduce environmental and societal resilience.

Both the traditional and western world consider technology to be an important and necessary tool for improving society's livelihood and therefore enhancing human survival. The tribes were, and are, readily adopting new technologies from bows to rifles, to computers, cell phones, etc. But the ethics of how and why tribal members use each technology is still constant and relatively unchanged. Tribes do not view technology as being "evil"; but how one uses it can be "good" or "bad". This contrasts the western world thoughts on technology where it is either considered good or bad.

Unfortunately ensuring a balanced distribution of benefits and costs during economic development is hampered by the necessity of making trade-offs among competitive choices. Since scarcity of resources is real, making trade-offs is difficult. Competitive demands for resources appear to marginalize parts of society even when they own or live on the land where the resources are collected [5, 6]. If industrialized societies today continue their current approaches to resource overexploitation, they will only accelerate the rate at which indigenous peoples have the opportunities of their livelihoods impaired. Collapse is a realistic endpoint for some groups as resource scarcity rears its ugly head. We need to consider a different approach when making resource decisions. We contend that the local to regional cultures and traditions of Native Americans, as well as global indigenous communities, can provide the road map to help refine tools and practices needed to become a sustainable practitioner (see sections 5–8).

Today, scarcity is not a choice and technology by itself will not give us a better life. Climate change has hung a large dark cloud over making sustainable decisions [2, 7] and will further exacerbate resource scarcity for some. This again points out differences in western societies versus tribal and/or indigenous societies perspectives on nature. Most tribal and indigenous societies are more dependent upon their lands for sustenance and thus more directly sensitive to disturbances such as climate change. Hence, most indigenous communities are not waiting for these changes to happen but are planning for a future that includes the impacts of climate change. They cannot afford to wait for climate change impacts to occur and then figure out how to survive in a new environment. They are dependent upon healthy rivers and lands to provide subsistence and cultural resources. These groups also have fewer options when the lands and rivers are not healthy. They are already experiencing the impacts of record floods which make each community more vulnerable to climate change (e.g.,// nwifc.org; http:cifor.org). Climate change impacts the social and environmental resilience of communities who have few options to survive when their land's ability to deliver resources decreases.

Resource managers, including forest farmers and fishermen, and sellers of ecosystem services/products need to behave like holistic managers who have a strong vein of humanistic behavior, without solely making nature into an economic commodity. This behavior is found in indigenous communities with a nature-based culture. This is a daunting task for industrialized societies to mimic because of the need to integrate all parts of the ecosystem. We need to practice adaptive strategies and not assume that our economic models are adequate to make resource choices. No one ever said being sustainable was easy! If this was easy to practice, all societies would probably already be doing it.

In industrialized countries, resource managers and politicians still struggle to include humans equitably in the resource management process while protecting nature and its ecosystem services. This difficulty arises because **"Human Equity"** and **"Nature"** and **"Ecosystem Services" are considered separate entities**. If you understand that they are one and the same, becoming a sustainable practitioner is more tenable. In this book we introduce our ideas on how to link nature and society to make sustainable choices. Our contention is that to be sustainable, nature and its endowment needs to be linked to human behavior similar to the practices of indigenous peoples. Human behavior can be examined through a cultural lens such as by the stories that a group of people tell each other.

We are also using a water metaphor to provide insights to the complexity that surrounds a scarce resource and as a metaphor to describe the connections between human and nature as practiced by indigenous communities. This is also very appropriate since water is considered the first of First Nation foods. Native Americans' view on water is that all things come and go through water, both naturally and spiritually. Also "kush" (Nez Perce for water, and similar words in other related languages) means "amen" and is said at the end of prayers.

We want to thank Cal Mukumoto and Toral Patel-Weynand for numerous insightful discussions and being part of workshops that helped the authors to develop the themes of this book. We had many invaluable discussions with both individuals and they really helped us to think through our ideas and how we wanted to write this book. We couldn't have started writing this book without their discussions and thoughtful inputs. We also want to thank Myrna Tovey, John D Tovey's grandmother, and the many conversations that John had with her about the ideas written in this book.

Melody Starya Mobley (Cherokee) contributed considerably to the editing and writing of materials for this book. She is a Natural and Non-Renewable Resource Conservation Planner. She is a lifetime member of the National Congress of the American Indian; is included on Wall of Honor in the National Museum of Native Americans, Smithsonian Institution. Starya, Ms. Mobley's Native American name, is derived from the Cherokee language term "Stah-yu" which means "Stay strong."

The Authors
31 December 2012

# Contents

# List of Boxes

# Sustainability:
# Learning from the Past

# Chapter 1
# The Context for Our Sustainability Story

## 1.1 Post-1492: European Colonialism Impacts on Peoples of the Americas

We cannot cover in this book the multiple impacts of the European colonialists' practices on the peoples of the Americas. We do encourage the reader to go into the literature and read many excellent volumes of materials that have been written on different aspects of this topic. In this book we want to introduce the idea of how the American populations collapsed upon the arrival of the Europeans and will include some of the battles faced by Native Americans in section 2. We encourage the reader to read Charles Mann's excellent books written on this topic for the Americas.

When these European countries began their conquest more than 500 years ago, the lands were not vacant and many indigenous people had built highly complex societies. These societies were not uncivilized or undeveloped. However the European's still treated these indigenous people as the upriver people as we describe in our river metaphor (section 1.3). European colonialists did not care whether the civilizations they conquered were sophisticated or civilized since these peoples did not practice the European models of what it meant to be civilized. Of course, underlying all of this was the simple fact that Europeans wanted to exploit resources that did not belong to them for their own benefit. It was easy to justify taking someone else's resources if you could rationalize in your own mind that they were not "civilized" and you knew better how to use their resources. Of course, it didn't hurt that the Europeans also desperately needed these resources.

The civilizations conquered by the Europeans were sophisticated and highly complex societies that collapsed upon the arrival of the Europeans [8]. William R Fowler described how sophisticated and civilized the Inca Empire was when the Spaniards first arrived in his contribution to the Microsoft Encarta in 2000 [9]:

> "...The Incas built a wealthy and complex civilization that ruled more than 9 million people. The Inca system of government was among the most complex political organizations of any Native American people. Although the Incas lacked both a written language and the concept of the wheel, they accomplished feats of engineering that were unequaled elsewhere in the Americas. They built large stone structures without mortar and constructed suspension bridges and roads that crossed the steep mountain valleys of the Andes."

Initially these indigenous civilizations cautiously welcomed these "white people", e.g., the Spanish explorer Francisco Pizarro, and his 180 soldiers, when they landed in 1532 on the Peruvian coast [9]. The Incas also had a prophecy that linked a white person as being their God returning to this world. These were the first white people the Inca had met. Fowler wrote

> "...The Incas at first believed Pizarro to be their creator god Viracocha, just as the Aztecs of Mexico had associated the Spanish explorer Hernán Cortés with their god Quetzalcoatl..."

Spaniards were so god-like in their appearance that this perception was not unrealistic. The Inca had never seen horses and guns that the Spaniards brought with them. Large horses were an extraordinary sight. The horses even obeyed their Spanish riders. Spanish guns also produced a lot of smoke, made a lot of noise and killed so well.

By the time these civilizations realized that the European colonialists were not gods, it was too late. The dramatic transformation of the cultures and civilizations of North, Central and South Americans had begun.

The Europeans introduced new diseases to these lands which decimated indigenous populations. Millions of people died. For the Inca civilization, it took less than 30 years before these new diseases killed the rulers of the lands [9]:

> " About 1525 both Huayna Capac and his appointed heir died ... probably from one of the European diseases that accompanied the arrival of the Spaniards..."

In central Mexico, the arrival of the Spaniards introduced many new diseases and significantly reduced this region's population [10]. Mann [18] summarized how the population in central Mexico decreased from about 25.2 million in 1518 to 6.3 million by 1545 and finally to 0.7 million by 1623. In only 30 years, only a quarter of the population of central Mexico was still alive. These population losses were the result of diseases like smallpox, measles, cocoliziti, plaque, and influenza [10]. This story was repeated throughout the Americas and was an effective tool to remove the indigenous populations so the European explorers could take their lands [8].

Scholars have debated on how large the indigenous populations were upon the arrival of Columbus and other European explorers. The exact number is not as relevant as the fact that the population size was in the millions. Prior to the arrival of Europeans, large pre-1492 population densities and highly developed civilizations existed in the Americas.

These civilizations collapsed because of the European colonialist who wanted the resources found on these lands. The Europeans traveled from home countries that had already over-exploited their resources. Most of the European countries had already *refined their colonialism model* while competing for resources in Europe, Africa and parts of the Middle East. This will be briefly described next.

## 1.2 Post-1492: European Colonialism: Thirst for Resource-rich Lands

What stimulated the western European countries to expand their influence to such a far-reaching extent around the world? Resources scarcity contributed to the development of the European colonialism model. Resources scarcity has been a problem that European rulers have cyclically faced for more than a thousand years. Initially they settled the problem of resource scarcity by rulers designating most in-country lands and resources for the sole use by the crown. Crown lands were needed to provide the wood to build the royal navies and to fire the foundries to manufacture rifles and cannons. In-country resources were important to build a country's military power since forests were needed to build ships and wood heated foundries to manufacture weapons. If you had a strong navy and could manufacture weapons, it increased your chances of becoming an economic, military and political power house in Europe.

Eventually in-country resources were over-exploited and resource scarcity became a problem again. If rulers could not acquire the supplies they needed to build their economies, it reduced their economic, military and political status in Europe. This resources scarcity problem was not an acceptable situation to the European rulers since countries frequently waged war on one another. Any European country could potentially be conquered by some other European power.

The solution to this scarcity dilemma was to conquer other lands rich in resources, i.e., "colonialism model" of resource acquisition. By conquering other lands, the resources could be shipped back to their home country. Perhaps, even the indigenous peoples could be used as labor to collect the resources. This appeared to be a win-win situation for each European country. They could maintain their status while acquiring luxury items that were more scarce in Europe, e.g., spices, salt, timber and gold, to name a few. They could acquire resources to industrialize their economies but at the same time produce products with economic value in global markets.

Initially this colonialism model of conquest was successfully practiced by Europeans in locations close to the European continent. Historical records document how the Roman Empire over-exploited forests in North Africa to maintain their empire from 30 to 476 CE; they even set up a forest reserve in Africa for themselves when concerns were raised about shortages of timber [11]. By conquering other lands, European powers appeared to have resolved their scarcity issues but made resources scarce for the people they conquered. Eventually they needed to expand their territories to acquire more resources since too many European countries were competing for the same resources. Since the colonialism model worked so well, it was a logical path for them to continue practicing this but in more distant lands.

After 1492, European countries significantly expanded their conquests to parts of the world that they had not explored before. Several countries sent explorers to find new territories that could be conquered and to find new sources of resources. They eventually arrived on the shores of the Americas. Conquering North America was not easy for the European powers since they faced battles on two fronts. They fought indigenous communities for ownership of their lands but also fought other European countries who wanted the same lands. For North America, the dominant players in the fight to control access to conquered land resources were the British, French and Spanish [12]; after 1492 the only European countries that did not participate in the battles to own North America were the Scandinavian countries. This is not to say that the Scandinavian countries were not adventurous explorers. Before 1492, the Scandinavians were some of the world's premier seafarers searching for new lands, e.g., Leif Erickson and others.

For western European countries, access to resources from conquered countries allowed them to resolve their resources scarcity issues. These resources were used to build a country's military strength, e.g., navies and armaments, so they could dominate global trade routes. These countries would not have become global military and economic power houses had they been restricted to only consuming in-country resources.

In North America, the Europeans found an endless abundance of tradable resources that must have been a pleasant surprise to them. At first, the Native populations were often treated as trading partners and the royal charters usually gave licenses to the companies to conduct trade with the tribes. Additionally, both sides felt they were cheating the other. The drawback of the western cheat just happened to be more disastrous than what the Indians thought they were cheating out of the Europeans.

Ultimately, it was a resource "feast" for the European conquerors since they had landed in a geographic location where nature is highly productive [1, 13]. Fortunately for the first European colonialists, lands in North America are highly productive compared to many other regions in the world [1]. These lands provided abundant natural resources and the rich soils were good for agriculture. Nature had not been over-exploited by the Native Americans in North America. However Native Americans did not follow the European model of resource management [14] (see section 5.3).

How nature was treated changed when the European colonialists arrived on the American continent. Europeans continued their exploitive practice learned in "resource scarce" Europe. It did not take long for the east coast U.S. fisheries to collapse. The first documented case of fisheries collapse was already reported in the 1700s in the Atlantic coast of the Americas [15]. Despite the reported collapse of resources, European colonialists were able to enrich themselves and the coffers of their rulers who lived in Europe. Many ship loads of resources sailed back across the Atlantic Ocean to Europe.

In general, western European countries gave little thought to the indigenous communities' living and surviving from the lands when they began their global conquest. Those who conquer other people usually do not worry about whether they treat the conquered people well. Western European countries fit this mold and were only interested in resources and lands they could add to their possessions. They had a 500-year period of global exploration while industrializing. This long period of exploration ultimately dramatically changed the environments and societies in North America, South America, Africa, and Asia. They left no part of the world untouched during this period of exploration.

For reasons just described, resource practices of the colonists were not sustainable. These types of practices will continue to produce an inequitable distribution of resources where the buyers of resources continue to marginalize the growers or producers of resources. The industrialized countries would continue to gain most of the benefits of resources while the developing countries would continue to be the suppliers of resources. In this case, industrialized countries use western economic models to ensure the continued delivery of ecosystem services for their benefit.

If sustainability is the real goal, global societies need a different glue to implement sustainable practices. This need for a different glue or portfolio of practices will be briefly discussed in the next section. We need to begin thinking about what the management practices of the non-western world are and what kind of portfolio or suite of tools would result from changing the western world behavior towards nature.

## 1.3 We Need A Different Glue to Make Sustainability Work

*"Everything on the earth has a purpose, every disease an herb to cure it, and every person a mission. This is the Indian theory of existence."*

~ Mourning Dove (Christine Quintasket, Okanogan) ~

The colonialism model of global conquest, and over-exploitation of resources not belonging to them made these countries powerful and able to build global empires. It was decidedly unsustainable harvests of others' resources. Indigenous communities frequently collapsed upon losing the resources that were the bases of their livelihood and survival. Practicing over-exploitive harvest of indigenous community resources, and isolating these communities from their customary sources of resources cannot be considered sustainable, especially from the indigenous communities' context! It is not the glue that should be used to frame sustainability choices. We cannot make decisions that transfer the costs to future generations or to those more vulnerable to their environments because of unsustainable practices. We need to develop a different model of sustainability that does not build upon past colonialism models. We can get these models from native people who have survived for hundreds of years on their lands while dealing with a foreign intrusion on their cultures and traditions. The survivors of the people who live in the conquered lands have many stories to tell us. Foremost among them would be how to practice a non-western world model of human development.

We contend that there are many lessons to learn from a people who has lived for several thousand years on their land, and then lost those lands to foreign invaders. They adapted and survived to re-emerge as important drivers of natural resource policies today (see section 9). These are the people to learn from since they are superb adapters. They were able to retain their identities and cultures despite being conquered. We are talking about Native Americans and other indigenous communities that have strong, local, nature-based cultural roots. Culturally-based communities are unwilling to forgo certain traditions for purely economic gains. To pursue such a model would require a value change for the western world people.

We are not condoning an approach that is grounded in the false paradigms of the "savage" or "noble savage" [16]. Nor are we recommending society move back to a "Garden of Eden" and give up their technological and industrial achievements. This false paradigm is not sustainable. We have too many humans living on this globe to even consider moving back to a Garden of Eden. An apple can only be split up so many times before there is nothing left! No Native

American would suggest such an approach. This is a "cosmetic" change only when a behavioral change is needed. Native American Indians are not anti-technology. They have adapted to technology and can be found as studiously texting on their cell phone as anyone else. They don't send up smoke signals as only the most naive would think today after watching old western movies produced in the U.S.

How is it possible for Native Americans to be a model of sustainability if systematic efforts were made to eliminate all their cultures and traditions, i.e., to "civilize" them in the eyes of the European colonialists and later by the Euro-American settlers? We are focusing our story on the Native American tribes who have lost much of their land and customary resources. Even though Native Americans lost their lands resources upon the arrival of European colonialists to North America, today they provide leadership in natural resource management in many parts of the U.S. (see section 9). These people are not holding deep grudges that would stop them from contributing to restoring nature that was so highly altered with the arrival of the European colonialists. This means that there is a good story here, that is, global citizens could benefit from knowing if the world is truly invested in becoming sustainable. This is not a story of collapse [17], which human societies are drawn to read about, but a story of redemption of indigenous cultures that lost much with the arrival of the European colonialists.

The western European approaches to resource uses and development were disastrous on the environment and people they conquered. They introduced a "foreign" model of resource exploitation on the conquered lands. For sustainability to be successful it needs to be grounded in a "local" model of resource exploitation. "Foreign" models of resource consumption are grounded in an artificial paradigm since the local knowledge is not used to make resource decisions.

Local models of resource, practiced by indigenous communities, are still not understood by the western world. In general, sustainability is a "slippery slope" for the average western industrialized country citizen. They do not understand the problems and connectivity that exist in ecosystems because they are "foreigners" to the realities of the issues or problems they face. The average citizens in the industrialized countries are better at dealing with global and large-scale issues but not the "local" issues that need to be informed by knowledge collected over many centuries. Issues can range from mundane to local, e.g., what kind of toilet paper is "green" or "when does my septic system need to be emptied". Issues may even range to a larger scale such as critical foreign issues or book-learned problems such as how much deforestation is occurring, or is deforestation in Indonesia contributing to climate change, etc. A perfect example of this is described by a Chair of a Washington tribe when he mentioned how he frequently had to deal with mundane-to country-level problems as a normal part of his job. One day he had to help fix a septic system problem of a tribal member and then the next day fly to Washington D.C. to meet with the President of the

U.S. to discuss national tribal governance issues. In this case, local knowledge is being integrated with national knowledge to address a broad range of social and environmental problems. No problem is too mundane. They are all part of the essential leadership skills useful to those wanting to make sustainable decisions on resource uses and economic development.

We also need to look for insights from societies who learn inter-generational leadership skills and where traditional knowledge helps them to adapt to live within the bounds of the industrialized societies who conquered them. These are people who think and live bounded by traditional knowledge learned within their communities but are able to integrate knowledge from their conquerors to adapt to the altered world that they live in today. They have succeeded in a way that can teach the western world some lessons. Their tools are ethically grounded because they have not lost their culture-link to nature. They do not assume that technology or economic assessments will provide us all the answers to resolve the "ills" of today's societies. If you want to know why we think there is a need for an alternative parable, you need to read further in our book!

The western scientific approach to environmental problems has been to search for solutions after we have already lost something. This situation has played out numerous times throughout the history of the western world. Of course, the western world is searching for solutions to some very tricky problems, e.g., conservation, poverty reduction, climate change mitigation, etc. These problems have become tricky because the western world typically looks for solutions to a problem once the problem has already erupted. As Benjamin Franklin wrote many years ago [26]:

> "When the well's dry, we know the worth of water."

This means more attention is paid to its negative repercussions of the problem than the original causal factor that created the problem. It is also very difficult to resolve a problem when one is dealing with all the negative repercussions and impacted stakeholders. American tribes do not have this short-sided approach to resolving conservation, poverty, economic development and climate change problems. The Native American approach to environmental problems is to restore or mitigate problems before they lose something of cultural value. The efforts implemented by the Swinomish Tribe to restore salmon habitat are one example of the fore-sightedness of tribes to allocate funds towards resolving problems created by western world economic activities. Many tribes are implementing projects or planning to restore and mitigate habitat losses so salmon can return to streams (see section 9).

We think it is worthwhile to have a brief detour in our story to introduce our water metaphor. Why are we using a water focus, i.e., the river, to tell our story for this book? Water crosses political, industrial and land-ownership boundaries and is solar driven. A river has a beginning and an end. It runs in only one direction—from its source to its mouth. In between the headwaters and the river mouth, humans and nature play out their stories. These stories impact life from beginning to the mouth of the river but also beyond the riverscape. Rivers are found all over the world and generally no one group owns it. Socially and inequitably distributed water is a common theme found in our entire human history. So a river story is a global story and a metaphor for society and nature in space and time. Fresh water is a resource that is sacred, it is also scarce and many battles have been fought over who has access to it. It is an excellent metaphor for the western world and indigenous people's sustainability practices for a common resource that is difficult for a person to own.

## 1.4 Essential Sustainability: Insights from A Water Metaphor

### 1.4.1 Water—A Scarce Global Common Resource

Rivers have historically been the waste dumps or sewer ducts of industrialized societies giving rise to the common saying— "dilution is the solution to pollution". Hence rivers are also the repository of human diseases and disease vectors. There are many historical examples of people polluting rivers even though they are dependent on its water for food, transportation, energy and many other products. Humans typically ignore the "river's messages" telling them that the river is no longer able to tolerate and absorb additional human wastes.

Humans have commonly not treated their rivers well and the river has "bit back" on human health and survival. Even when people became sick from drinking river water, human dumping of wastes did not stop since it was generally only the lower- class citizens who were impacted by sicknesses. Political leaders were not worried about the diseases that were rampant in London during the mid-1800s. One reason was that they did not need to drink the water collected from the Thames River. Only the lower-class citizens drank from this contaminated water.

When people in power were finally directly impacted by the unhealthy river conditions, policies changed in to cleaning up the river. London began to deal seriously with dumping its sewage into the Thames River only after the river became excessively smelly. Baker [19] wrote

> "...the Thames River, which received most of London's 'new sewage', became so polluted from the new discharges of sewage that the stench

forced Parliament, located on the banks of the Thames, to adjourn! The 'Great Stink', as the 1858 event became known, motivated Parliament to provide the funding for an extensive renovation of its sewer system..."

**Fig. 1.1** A water metaphor for how water links nature and urban areas. Figure credit: Ryan Rosendal.

Around the world, rivers have provided a dual function for humans. They have been the transportation life-blood of many large cities and at the same time the repository of the wastes that humans throw out. Humans have altered the flow of rivers to suit their needs from the development of wells, irrigation canals and aqueducts to modern hydro-electric dams. When industrializing, river channels were straightened to facilitate their use as a transportation network. Humans have even split continents using water, taking the Panama and Suez Canals for example. We have reduced its meander and channeled it when it travels through human-built environments because of flooding concerns, even though later understanding reveals this actually exacerbates flood events. Technology is used to extract water from deep in the earth for industries and for agriculture, and to dam a river to generate power.

Rivers also have been the vehicle for humans to make money if you could control access to who could travel it. The Anonymous saying—*Water flows uphill towards money*—captures well how controlling water access has been important for societies [20]. There are many stories from Europe of nobles or chieftains demanding tolls from boat captains to pass up or down a river to travel to other cities to trade their products.

At the same time, fresh water has connected, and continues to connect, humans to nature. Many groups have protected water by establishing sacred sites to ensure a community's access to sufficient clean drinking water. Sacred areas established around the world frequently protected soil and water, e.g. in Japan in the 1500s, in Switzerland and Austria in the 1800s [21].

### 1.4.2 Water as A Sacred Resource

The fresh water that flows in rivers has been a focus of attention of many enlightened people living in traditional as well as industrialized societies. People recognized the values and roles of water and the creatures that lived in the water that humans ate to survive.

A few proverbs highlight how enlightened thinkers recognized the importance of water and how strongly humans are connected to it (www.cyber-nook.com/water/p-quotes.htm):

- In every glass of water we drink, some of the water has already passed through fishes, trees, bacteria, worms in the soil, and many other organisms, including people. — Elliot A. Norse
- Water is the driver of Nature. — Leonardo da Vinci
- I never drink water because of the disgusting things that fish do in it. — W.C. Fields
- Plans to protect air and water, wilderness and wildlife are in fact plans to protect man. — Stewart Udall
- Ocean: A body of water occupying two-thirds of a world made for man— who has no gills. — Ambrose Bierce
- I believe that water is the only drink for a wise man. — Henry David Thoreau
- Pure water is the world's first and foremost medicine. — Slovakian Proverb

Water is the basic fluid essential to all human survival. Water appears in many cultural and traditional ceremonies. It is not unusual to find societies managing traditional sacred areas mention of water nymphs or spirits living in and protecting water.

Water continues to be an important part of Native American culture and traditions today (see Box 1). Fresh or non-salty water is strongly linked to indigenous cultures due to its historical scarcity. Water is still drunk during ceremonial dinners as described in Box 1.

**Box 1: Drinking Water before Ceremonial Dinners—Native American**

As a manner of respect, many tribes in the Plateau Cultural areas of Washington, Oregon and Idaho begin and end their ceremonial dinners with drinking water. A person is selected to serve the water and pour a small amount into each cup at each setting. Water is served before any other food is placed on the table and it is important that this person have prayerful thoughts and a good heart. At some of the tribal longhouses, religious songs will be sung while the water and other traditional foods are served. One example is the root feast which is held in the springtime each year when edible roots such as white and black camas, wild potatoes, bitter roots and koush are harvested and everyone gathers at their longhouse for this first foods celebration. Other traditional foods are served at this dinner such as deer meat, salmon and berries. This dinner is an honoring and appreciation for these foods and it is a time to give thanks to the Creator for bringing these important resources back. Many Plateau tribes believe that every living being on the earth follows its own natural laws, the cycle of life. Water is important to every living thing on earth. It's the way things are and if there was no water there wouldn't be no roots, salmon or berries and no cycle, no food. After singing songs of prayer for the foods and water, the meal begins with a small sip of water. These foods and water are considered sacred. This is a festive occasion and everyone enjoys the company seated together. Everyone remains seated until the conclusion of this meal and this will be signified with a final sip of water. Once the dinner is completed, nothing is wasted and guests will gather what food is left and take this home.

In general, all indigenous people respect water and water is an important part of their ceremonial activities. The Sundanese tribe in West Java and Banten provinces, Indonesia practice water rituals. The circumstances and the ceremonies followed by the Sundanese tribe vary from the practices followed by Native Americans but the respect for water is the same. For example, during the Seren Tahun ceremony, holy water (air suci) is collected from seven river headwaters and is sprinkled on every single person who attends the ceremony (see Box 2).

In Indonesia, the Sundanese tribe does not have a word for "hungry" in their vocabulary since water provides them abundant food as long as they practice their rituals and respect for water. In contrast, they have plenty of expressions to describe a "full stomach", such as "seubeuh", "wareg", "bentet", and the superlative "kamerekaan" [22]. The Sundanese are blessed by having access to plentiful supplies of water from rivers. This water is used by the community to easily plant rice in their paddy field and vegetables in their gardens. Raising local fish at a fish pond and cattle in their home garden are not difficult tasks since water makes everything possible.

**Box 2: Waterfall Ritual at the Fourth Congress of Indigenous Peoples Alliance of Archipelago—AMAN**

A water ceremony was conducted at the official opening of the fourth Congress of Indigenous Peoples Alliance of Archipelago (Aliansi Masyarakat Adat Nusantara/AMAN) on April 23–24, 2012 in Tobelo, North Halmahera district, North Maluku province, Indonesia. The ceremony was located at the Water Monument of Tobelo where 1,669 tribal peoples across Indonesian Archipelago—who brought sacred fresh water from their villages—poured the water into the Monument of Water. The sacred water is collected from sacred places across the Indonesian archipelago. It is stored in special containers like bamboos, jars, bottles, and other vessels that correspond to the local values of indigenous peoples in each region.

In 20–21 April AMAN also facilitated a culinary festival that serves traditional cuisine representing traditional cultures and natural resources of AMAN members. Most of the cuisines are not rice-based foods indicating that indigenous peoples in Indonesia are not relying on rice—the major staple that has been promoted by the previous Indonesian government, particularly during the Green Revolution.

In the Island of Bali, during purification rituals to welcome Nyepi—a day of silence, fasting and meditation—the Balinese people flock to beaches in this Paradise Island. They carry colorful offerings and perform Pratima (sacred effigies). Padang Galak Beach, five kilometers from the Bali province's capital, becomes one among many sacred beaches used for the Melasti ritual. This ritual aims to purify sacred objects belonging to several temples, and also to acquire sacred water from the sea. After performing a sacrifice, Balinese bring the pratima to the coast, as well as dipping their feet in the sea water. Other Balinese who cannot attend the ceremony will be sprinkled with seawater collected by their family members.

(Source: various readings, AMAN website)

This connection of most indigenous communities to water is apparent from the numerous Native American and other indigenous village sites that are always located along rivers and fresh water drainages. In Sundanese, it is easy to find a name of a city or village that begins with the word "water", e.g. District Ciamis (sweet water), Sub-district Cibodas (white water), and Cipanas (hot water) village. Clean and cool water is a critical resource important to the livelihood and cultural survival of many Native peoples. Native American people use resources for sustenance that live in water, e.g., salmon, at a much higher consumption rate than any global people.

The sacredness of water meant that community taboos were used to control and to stop over-exploitation of water. Communities guarded these areas. Native Americans have many stories to teach children on the value of water and how it should not be abused. One Native American proverb states this well:

> "The frog does not drink up the pond in which he lives."

Native Americans have knowledge and a deep relationship with their natural environments that has developed over many generations. Climate change and massive population growth, which will impact habitat loss and increased use of water resources, will affect Native Americans and indigenous communities more than anyone else because of their cultural and daily link with rivers, lakes and oceans.

### 1.4.3 Water, Water Everywhere but Still Scarce

No one argues that without water, human life on the earth would not exist. This quote published by the United Nations [23] summarizes the importance of water for societies:

> "Without food we can survive weeks. But without water, we can die of dehydration in as little as two days."

Water is a scarce resource that many governments have not been able to provide its citizens. The 2012 UN report states [23]:

> "Water is our most valuable natural resource. It is essential to all basic human needs, including food, drinking water, sanitation, health, energy and shelter. Its proper management is the most pressing natural resource challenge of all. Without water we have no society, no economy, no culture, no life. By its very nature and multiple uses, water is a complex subject. Although water is a global issue, the problems and solutions are often highly localized." ··· more than 40% of the world's population, experience some form of water scarcity ···

Human life would be impossible without fresh water which is why so many wars have been fought to own it or control its supply. Many treaties have been written to share and distribute water. In 1993, Gleick summarized the history of global water conflicts dating back to 3000 BCE [24]. During this 5,000 year history, 186 major battles over water occurred in every region of the world. Over half (52%) of these conflicts were part of military operations while another 31% were labeled as terrorism activities [24].

During battles, denying people access to fresh water becomes an effective weapon of war. When visiting castles in Europe, it is common to find a deep well built inside every castle wall. When fighting erupted, a castle had to be able to close its gates and fight the attacking army from inside its fortifications. They also had to be able to survive for many days inside the castle walls during a siege.

Access to fresh water has been a problem throughout recorded human history. It continues to be one of the really "tricky" problems that society faces today. We need to develop equitable and sustainable solutions for delivering water to society. If we cannot make this happen, conflicts or battles will continue unabated over this increasingly scarce but essential resource.

Industrialization and feeding a large population are responsible for much of the water scarcity that we face today. According to a 2012 United Nations report [23]:

> "Agriculture, industry and energy are the biggest users of water—it can take 10,000 litres to produce a single hamburger, 1,000 to 4,000 litres for one litre of biofuel and 230,000 litres for a tonne of steel. Agriculture alone accounts for 70% of water use worldwide."

This latter quote explains why societies expend considerable financial resources to build large-scale infrastructures to manage water supplies. Building water management infrastructures highlight the competing demands for fresh water supplies. The builders of these infrastructures determine who has access to water and its ecosystem services. This means that water is a good medium to look at how we make sustainable choices since access to fresh water has costs and benefits, and frequently someone always seems to be a loser. It highlights the inequities that exist globally on the benefits and costs to accessing fresh water, and the competing interests to control where it flows and who benefits from it. Indigenous communities, including Native Americans (see section 2.2.4), provide many examples of people who lost much when dams were built on their lands for the benefit of "down-river" people.

Equitable access to fresh water is one of those tricky problems that society faces since it pits people and their cultures against economic development and financial rewards. Up-river people seldom see the benefits of the rivers that originate where they live. People living at the beginning of the river systems do not have much political clout since their population densities are generally low. Up-river people live in the watersheds where the forests grow that capture and store fresh water; "at least one-third of the world's largest cities obtain a significant portion of their drinking water directly from forested protected areas" [23].

So as a recap, down-river and city people acquire most of the benefits when dams are built, e.g., electricity from hydro-electric dams. Down-river people build the infrastructures in the up-river locations to control their access to these distant water supplies. These down-river people live in cities with larger population densities. They have the political power to make regulations to determine who and how much water will be consumed by whom.

Who controls water supplies is not just a problem for up-river and down-river people living in one watershed. This is also a common source of conflict between countries since rivers meander and cross many country borders; few countries own and totally control their fresh water supplies [25]:

> "Many rivers, lakes, and groundwater aquifers are shared by two or more nations, and most of the available freshwater of the Earth crosses political borders. International watersheds cover about half of the Earth's land surface, and about 40% of the world's population relies on these shared water sources."

Political boundaries rarely coincide with borders of regional watersheds so politics and wars predictably will decide who controls water rights and access. Gleick [24] reported that over 260 river basins and ~270 groundwater aquifers are a resource owned by at least two or more nations. He further estimated that 60% of the global fresh water is a shared resource between at least two countries [24]. The Jordan River Basin, for example, is managed collaboratively by several countries since no one country owns the total river. The Jordan River originates in the Jordan Basin and treaties determine how much water should flow to Israel; 40% of Israel's fresh water derives from this source [26]. Even though down-river people benefit the most from river water (unless too little water reaches down-river or what does arrive down-river is very salty), the infrastructure and technology to manage water supplies are typically located at the source of the water, i.e., up-river locations. Both Israel and Jordan have built irrigation canals or dams in the Jordan Valley to store water for irrigation.

The natural sources and channels of water change with time and therefore people only temporarily own water. Aquatic lands owned by the State of Washington were based on maps drawn at the time of statehood in 1889. The location of these aquatic lands has changed over time. Some of the shorelines and river beds have changed and the state now owns some dry river channels and submerged ocean shorelines. This makes the ownership of water ephemeral. Water scarcity is therefore a reality that changes with time and place.

Water sources today mostly come from surface water, ground water and even the ocean. However for this book we will mostly discuss surface water sources to parallel the river metaphor used in this book. In fact the industrialized world consumer's knowledge of water production and supplies is abysmal. Many water consumers living in cities in industrialized countries think water comes from the faucet in their kitchen. These water consumers derive all the benefits of the water without recognizing that there are costs and benefits to people living at the source of the water (up-river in this metaphor). This inequitable distribution of resources and alteration of nature to increase society's access to vital resources are exemplified by the western European colonialism model of expansion and even conquest. The colonialism model had winners and many losers as natural resources were over-exploited and local areas of scarcity occurred. These patterns of colonialism highlight the practices that were not sustainable in the past and are still not locally sustainable today. These colonial models transformed regional societies and cultures when they imposed their brand of resource use. These cultures had already adapted to their environments over several hundred if not thousands of years. However it did not take long for regional cultures to lose the ability to practice their traditions.

In addition to our water metaphor, we want to introduce our "coyote" mascot because it reflects the multiple facets of sustainability. The coyote is considered a central figure in several Native American cultures (see Fig. 1.2). Tribal stories, like the coyote stories, have a focus to not only entertain but also to teach a moral [27]. The coyote is nicknamed the "trickster". He is neither "all good" nor "all bad". He exhibits both the good and the bad characteristics of human beings. The coyote is relevant since many "sustainability practitioners" think they are only a force for good. Stakeholder groups, however, have biases that they need to recognize. Stakeholder groups are not coyote-like and mistakenly think that each decision made is either good or bad. They do not see that each decision has good and bad elements to it or even shades of gray are present in each decision. We need to think more like a coyote if we are going to make more sustainable decisions for natural resources. This means we should not polarize society into special interest groups that do not accept knowledge held by those not in the inside group. Again we mistakenly think that only one of those special interest groups is correct rather than considering that perhaps "truth" may lay within several groups. We need to recognize that no one has sole ownership of the correct knowledge or the truth. We hope you understand the conceptual meaning of our mascot and appreciate its relevance!

**Fig. 1.2** An image of a Native American coyote.
Illustration credit: Ryan Rosendal.

## 1.5 Our Coyote Mascot Blends the Dual Nature of Sustainability

The purpose of this book is to help us become more "coyote-like"—a worthy goal in our world today. Because the good and bad characteristics of society are embedded in the coyote, it is the mascot for our book. We will use the coyote because of its dual nature. It helps us to make transparent how tribes use different decision models compared to the reliance of the scientific approach of western industrialized countries. The coyote stories are just one of the many legendary stories told by tribes and are particularly suitable for our theme.

In Native American stories, the coyote was responsible for preparing native peoples for the arrival of white humans [27]. So it is appropriate that we introduce "coyote essentials" at the end of each book section that will inform the reader how to behave more Indian-like (at least in the sustainable resource-use context). Again the coyote is a symbolism that there is not just one approach to resolve a problem or just one solution. For example, no one is completely bad or good (e.g., completely a devil or an angel). The following Hopi proverb characterizes the message that the coyote expresses (www.rodneyohebsion.com/hopi.htm):

"If two different bowls both get the job done, then what difference does it make if one bowl is dark and the other is pale?"

The extract in Box 3 by Roberta L. Hall enlightens us on the dual nature of the coyote.

**Box 3: The Coquille Indians: Yesterday, Today and Tomorrow**

"First there are the 'Coyote' stories, mythic stories of a distant past in which animals and humans are of the same reality. The personage Coyote is multifaceted. To some people and in some tales he is the mythical creator of the human landscape, the one responsible for the world as the Coquilles knew it, and was responsible for humans and their role in the world. Coyote also is a trickster, a wily individual with some attributes of the animal 'coyote'. He is a clever being who tricks others and is tricked by himself; often he is both more clever than most of those who he encounters and, at times, more fallible. To some people, Coyote represents the devil. Other Coquilles believe that the original view of Coyote did not encompass the devil role, but that Coyote was refashioned as a devil by Indians who had been converted to Christianity."

(from Hall [28])

Before we write about how the European colonialists battled to eliminate Native American cultures and traditions, it is worthwhile introducing a tribal perspective on their beliefs and practices. This will begin to highlight how the Native American beliefs and behavior differ from the western world. This is a thread that will be explored several times throughout this book.

## 1.6 A Tribal Perspective on Sustainability[1]

On December 10, 1992, Thomas Banyacya, a Hopi spiritual leader and elder made a speech at the United Nations

"...many powerful technologies (will) be developed and abused... Nature itself does not speak with a voice that can be easily understood...who in this world can speak for nature and the spiritual

1 By Wendell George, Confederated Tribes of the Colville. Books written by Wendell George: *Coyote Finishes the People* (2011) and *Last Chief Standing: A Tale of Two Cultures* (2012).

energy that creates and flows through all life? In every continent are human beings that are alike but who have not separated themselves from the land and from nature. It is through their voice that Nature can speak to us. The rock drawing shows the Hopi prophecy:

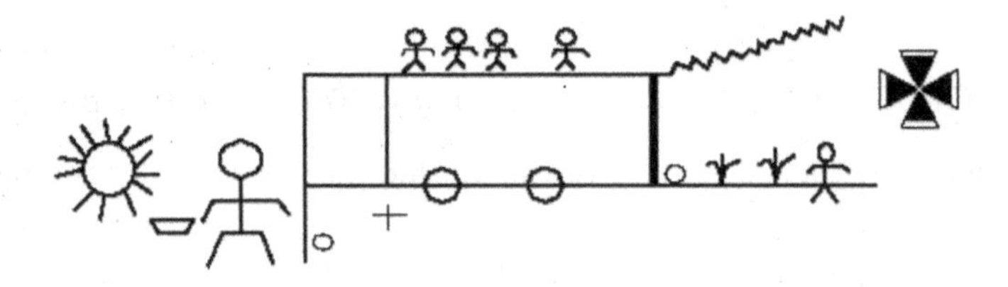

There are two paths. The first path has technology but is separated from natural and spiritual law and leads to the jagged lines representing chaos. The lower path is one that remains in harmony with natural law. Here a line is drawn that represents a choice like a bridge joining the paths. If people return to spiritual harmony and live from their hearts, they can experience a paradise in this world. If they continue only on this upper path, they will come to destruction. It's up to everyone, as children of Mother Earth, to clean up this mess before it's too late."

The idea of a Mother Earth was widely accepted as there was a Greek Goddess, Chaos, who gave births to the universe including the Goddess of Earth (Gaia).

But technologies are not all bad. Telephones, radio, TV and computers have made it possible for people to communicate with each other and expand their horizons. Ships, trains, autos, and airplanes do the same at a different level. All this is good if it ended there. But some people have used this technology to destructive ends such as war.

The issue can be reduced to the intent of the user of technology. The inventor or developer of the technology should be commended for his innovation and service to his community. But the people who put the technology to use without considering its impact on others should be condemned. American Indians try to prevent negative impact on seven succeeding generations.

However, technology has become so complicated and sophisticated that it is difficult to analyze. But most people recognize that we are economically gridlocked. We have allowed our technology to be used against us. This is leading us back to total chaos.

The only way to find the bridge that leads to spiritual harmony is to make a paradigm shift from the physical domain to the spiritual domain. This is like the Greek legend where Gaia evolved from Chaos. Gaia was the Greek Goddess of Earth. The Hindu's called her Kali and Native Americans called her Mother

Earth. All thought Gaia is the reciprocal interaction between everything on the Earth and the Sun.

One way to achieve harmony with nature is to relate our life to the Medicine Wheel. The Wheel is based on the sacred number four. The first four numbers of the universe are imbedded in music, astronomy, geography, and metaphysics. They make up the basis for the whole physical universe. For example, there are four elements of geometry: The Point, the Line, the Surface, and the Solid. (Also the 4th dimensions of the observable universe, 1st, 2nd, 3rd dimension and Time.)

These same four numbers are used by particle physicists to define the fundamental forces of the universe:

- An attractive or gravitational function: North on the Medicine Wheel.
- A radiative or electromagnetic function: East on the Medicine Wheel.
- A receptive or psychoactive function: South on the Medicine Wheel.
- A transmitting or informational function: West on the Medicine Wheel.

These four integers indicate unity of the psyche and matter. Today we are in a four-dimensional continuum: three-space dimensions and one time. The four winds described in many Indian legends are the same as the four directions of the Medicine Wheel except the wind indicates movement or something happening. It is related to many spiritual beliefs of the rest of the world.

For example, Chapter Two of the Acts of the Apostles in the Bible mentions a strong driving wind which in Hebrew (Ru-Ah) means "spirit, breath or wind". In Greek the word is pneuma, which means power. Indian stories say the North Wind is the strongest or most powerful in the physical world but each direction has a special meaning. South Wind is supportive. East Wind provides energy and West Wind is the intellect. This is the same spirit as what St. Paul said, which is poured in and out of our hearts. Indians always try to talk from their hearts. In Sura 15 (Al Hijr or the Rocky Tract) the Quran mentions the fecundating winds that cause the rain to descend from the sky, therewith providing water for humankind.

The Medicine Wheel is a powerful metaphor for the totality of life. All aspects of creation and consciousness, inclusive of the mineral, plant, animal, human and spirit realms, are contained within the center and four directions of the Medicine Wheel. They overlap and interweave to form the whole. It is in the center of the Medicine Wheel that we find the void, black hole, sacred zero, the chaos at the course of creation, containing all possibilities. Each of the elements—Earth, Water, Fire, and Air—is guided and molded by the sacred life force energy contained within the void. It is the source of chi, which is the life force energy. Life could not exist without the life-force energy of the void, which is the catalyst for all the powers that are found within the 360 degrees of the Medicine Wheel.

For American Indians it is a symbol of all creation, of all races of people, birds, fish, animals, trees and stones. The circular shape of the wheel describes the movements of the earth, the sun, the moon, and the cycles of life, the seasons, and day from night (unfortunately, it is not complete because the circular shape is two-dimensional and the universe is three-dimensional). This can be somewhat overcome if you look at the earth from the North Pole where you can see the counterclockwise rotational movement. This is the same as following the perimeter of the Medicine Wheel. At the center of the Wheel is the Creator who sits in perfect balance and harmony. Outside the center, there is the Mother Earth, Father Sun, Grandmother Moon, Grandfather Galaxy and the Four Winds or Directions. These represent spiritual paths, leading us to the center so we can also achieve perfect balance.

Scientists in the physical domain are reluctant to consider the spiritual domain. But this might be changing as Sir Arthur C. Clark said in 2006: "The future is not to be forecast but created".

The Hopi Indians predicted 200 years ago the world would be destroyed if we let our technology continue to lead us down the wrong path. This agrees with the Mayan prediction recorded about 670AD which said the "Age of the Jaguar" will end at the Winter Solstice on December 22, 2012. That will be the "gateway" to a new epoch of planetary development, with a radically different kind of consciousness.

Scientists such as Ervin Laszlo (*The Chaos Point*, 2006) and Steven Strogatz (*Sync*, 2003) agree with the Hopi that we are at a decision point. But they say our world is supersensitive so that even small fluctuations produce large-scale effects, hence the butterfly effect. The butterfly effect is demonstrated by a monarch butterfly flapping its wings in California and creating a tiny air fluctuation that is amplified and ends by creating a storm over Mongolia.

Laszlo also said we will reach the Chaos Point in 2012. He said, "The processes initiated at the dawn of the nineteenth century and accelerating since the 1960s build inevitably toward a decision-window and then toward a critical threshold of no return: the chaos point. Now a simple rule holds: We cannot stand still, we cannot go back, so we must keep moving. There are alternative ways we can move forward in. There is a path to breakdown, as well as a path to a new world". The new world will take us from an overly competitive environment to a more compassionate shared values community.

British scientist James Lovelock described the earth in his Gaia hypothesis as a living system that keeps it fit for life. He now thinks the system has been destroyed and may prove fatal for humanity. But that does not consider the impact of the human factor. Humans are capable of creating a butterfly effect.

A Mayan elder said, "The world will change for the better if we think with our heart rather than our head." Indians try to speak from the heart to be truthful and accurate. If enough people do this a new world will emerge just like a butterfly from a cocoon. When we become spiritually aware, we will be totally in control of our destiny.

One can already glean from George's perspective how Native Americans have not lost their link to nature and their cultures. Even though western European colonial countries attempted to destroy the cultures and ways of life for many indigenous communities, starting as early as 1200 and going to about 1700, these people survived without becoming part of the U.S. melting pot (see section 11.1). Instead, Native Americans became part of the "salad bowl" as discussed in section 11.1. Most societies would have become part of the conquering nation with only fading memories of what their life was like before this happened. This did not happen to Native Americans and is the reason their story can teach us about why traditions and cultures connected to nature make us more sustainable.

In the next section, we will briefly summarize how dramatically the European, especially English, colonialists impacted the indigenous communities living in present-day United States. This topic is important to cover since our hypothesis is that culture and traditions are the glue that keeps societies working towards resilient and equitable decisions. Once you understand all that the Native Americans faced and their ability to survive these impacts, the importance of culture to survive a war without losing your community values is extraordinary. This brief overview highlights how these colonialists recognized and targeted the elimination of Native American culture to control their resources and take ownership of their lands. It is important to understand what happened to Native Americans upon the arrival of the western European colonizers since these are not just old stories. The same approach is being implemented today by countries who think they are doing good deeds, e.g., establishing parks on indigenous lands to protect biodiversity at the continuing expense of indigenous communities who are evicted from their lands and are losing their traditional cultures. In 2011, farmers in Lampung and South Sumatra were killed as agrarian conflicts increased between peasants and plantation owners.

**Coyote Essentials**

- Maintaining ones cultural link to nature is the key to developing and implementing sustainable natural resource management practices.
- Resource scarcity always exists somewhere in the world.
- Colonialism models are typically/routinely used by those who do not own the resources.
- Only practicing good or bad behavior is not sustainable.

# Chapter 2

# Battles to Eliminate Native American Traditions and Cultures

> *"They never stop to get any kind of facts, and even if they hear, like Indians being half-hungry for a hundred years, no jobs, no homes, a lot of them say, 'So what, my people came from Norway. My people came from Germany.' 'They got here and they didn't have nothing, no land, no home, no nothing, but they got jobs, they went to work, saved their money, they worked hard.' Which they did, but they had something mighty different than the Indians, that was opportunity. The freedom of choice, all of the guarantees really in the first amendment. The freedom of speech, of the press, rights to gather in meetings and assemblies. To present their grievances to the government. Their right for redress, none of these things were given to the Indians. In our case, we're nine-years-old and we're beaten, and we're whipped for speaking our language. There was no freedom of speech."*
>
> ~ Quotes from the Harriet Dove's oral history ~

Who would be better to learn about sustainability than the people who survived having their cultures and way of life threatened, eliminated and transformed over a span of several hundred years after the arrival of European colonialists to North America? We are talking about over 500 Native American tribes who lived and flourished on these lands before the western European colonialists arrived on the shores of North America. In the proverb "das Kind mit dem Bad ausschütten" (see section 6.1), the Native Americans were the "baby" that was thrown out even though the original colonialists were dependent upon the good will of the native peoples to survive and were poorly adapted to live on these lands [12]. It is ironic that the symbolisms of today's land management agencies show the animals—eagles and buffalo—that are so revered by Native Americans. Similar stories can be found throughout the world. History has documented how it was the norm for more powerful countries to displace, rob and disenfranchise indigenous communities in their search for valuable resources.

## 2.1 European Colonial "Manifest Destiny"

The European Colonialists had many reasons for exploiting the resources they found in the "new world". They felt entitled, e.g., it was their "Manifest Destiny" or religious duty, to civilize peoples they conquered. At the same time, it didn't hurt that by exploiting resources from these conquered lands, they were able to maintain their imperial status and economic power against other European nations who were competing with them for the same resources.

At this period in history, forest resources were an important trade commodity. Trees were needed to build ships for the imperial Navy and commercial ships were needed to transport trade items back to Europe. In these days, ships were built out of wood. Unfortunately for the European conquerors, wooden ships were attacked by a mollusk, i.e., ship worm, which decreased the life of a ship. The British Empire was ecstatic when they discovered teak (i.e., a tree species in which its wood is resistant to the ship worms) in their newly conquered lands in India [29]. Wood was also needed to produce the charcoal used to heat metals in the foundries to manufacture cannons and other weapons of war. This meant a forest was cut only for the wood that grew in it. With tree supplies becoming scarce, this meant there was a high demand for trees by these flourishing industries.

When colonialists cut and cleared forests, their need and values supported the correctness of their actions. These values justified the over-exploitation of all resources on conquered lands. They felt justified to cut trees from lands that they really did not own and had no rights to exploit. Some of the reasons for their actions were:

- Progress—forests had no value for someone who wanted to farm the lands and farming was a civilized activity to pursue;
- Redemption—manifest destiny; and
- Fears—these lands were scary places that needed to be tamed and where strange wild animals needed to be killed since humans are a good food source for them.

European colonialists needed to clear land to also demonstrate their new ownership of lands that were already occupied by other people. They also wanted the lands to look just like what they had left behind in the "old country". These lands were also supposed to fit their idealized view of what nature was supposed to look like (see section 5.2).

### 2.1.1 Taming Indian Lands through Agriculture

Many of the English colonialists and settlers felt justified in their treatment of the indigenous peoples and taking their lands and resources. The lands appeared to be unused or vacant. European colonialists felt vacant lands needed to be used. It was not civilized to leave a piece of land untouched. They did not recognize that many of these lands appeared vacant because Native Americans made seasonal uses of them. Native Americans do not follow the European model of land-use or property rights (see section 5.3.1). Once the European colonialists transformed the land, i.e., clearing it, it was akin to heaven on earth. It couldn't get better for a Euro-American settler than to have a farm where a forest used to grow! It was their "Manifest Destiny" to civilize these "savage" people and the lands they had conquered.

Progress, under a Manifest Destiny umbrella, meant cutting down forests and converting these lands to farms. Since the English colonialists felt that the Native Americans were not using these lands anyway, they felt totally justified to make sure that every Euro-American settler had a piece of land that they could plow and own.

Thomas Jefferson's view on how agriculture and "being civilized" are integrally linked can be found in his writings to John Jay, James Madison, and George Washington [30]:

> "I think our governments will remain virtuous for many centuries; as long as they are chiefly agricultural; and this will be as long as there shall be vacant lands in any part of America. When they get piled upon one another in large cities, as in Europe, they will become corrupt as in Europe." (wrote to George Washington, December 20, 1787)
>
> "Cultivators of the earth are the most valuable citizens. They are the most vigorous, the most independant, the most virtuous, and they are tied to their country and wedded to it's liberty and interests by the most lasting bands." (wrote to John Jay, August 23, 1785)
>
> "It is not too soon to provide by every possible means that as few as possible shall be without a little portion of land. The small landholders are the most precious part of a state." (wrote to James Madison, October 28, 1785)
>
> "Good husbandry with us consists in abandoning Indian corn and tobacco, tending small grain, some red clover following, and endeavor-

ing to have, while the lands are at rest, a spontaneous cover of white clover. I do not present this as a culture judicious in itself, but as good in comparison with what most people there pursue." (wrote to George Washington, June 28, 1793)

### 2.1.2 Euro-Americans Settling the "Wild West"

Policies started by the European rulers, who funded these expeditions of conquest, were continued by the descendants of the first European colonialists. After the original settlers arrived and transformed the eastern parts of the U.S., their eyes turned towards the "Wild West" to control and also civilize these lands. Colonizing the west was a logical decision for the new U.S. government to make since they needed settlers to live on these lands to maintain their ownership of them. The Spanish, French and the British also wanted ownership of these same lands for themselves. The U.S. federal government also wanted their settlers to move into the dispossessed lands that used to belong to the Native Americans. It also did not hurt that gold was discovered in California in 1848. Many non-Indian peoples had dreams of free gold and wanted to travel west to go for gold prospecting to make their fortunes.

The federal government wanted its citizens to move out west, but it was not an easy task to stimulate settlement of these "wild" and unknown lands. Indians and wild animals still lived on these lands! The lands were not tame as they were in Europe. This made traveling west a potentially terrifying experience. You could die. The federal government needed to overcome people's fears of the Wild West and to view these lands as harmless. At the same time, they needed to stimulate a westward expansion of settlers for progress and redemption to maintain their ownership of these lands.

The federal government did not have money to fund infrastructure or economic development opportunities for the newly forming states joining the Union. What they did have was a lot of land in the public domain, e.g., all the lands taken from American Indians. Since collecting resources from public domain lands could be sold to generate revenue, this was a viable funding source to settle and develop the western U.S. The period of disposal of U.S. federal domain lands lasted from about 1782 until 1860. During this time, the U.S. government gave land away as homesteads and to the railroad companies to build the transportation infrastructure to move people westward.

The federal government gave public domain lands to settlers, companies and local governments. Homesteaders acquired ownership of lands if they managed them for a few years. The Homestead Act of 1862 allowed homesteaders to claim, manage and own any "unoccupied public lands". The Homestead Act allowed any person older than 21 years of age, "citizen of the United States, or who shall have filed his declaration of intention to become such, as required by the naturalization laws of the United States, and who has never borne arms against the United States Government or given aid and comfort to its enemies" to file claim to land to not exceed 65 ha (www.pbs.org/weta/thewest/resources/archives/five/homestd.htm).

The Homestead Act of 1862 was followed by several other Acts—all with a goal of allocating land to U.S. citizens. For example, the Timber Culture Act of 1873 allocated 65 ha of land to anyone who would start planting trees on 16 ha of the total land area they were allocated.

It also provided railroad companies land grants and right of ways to expand the transportation infrastructure from the eastern to the western reaches of the U.S. Railroad companies competed to build this network of railroad tracks. Railroad companies began to transport Euro-American settlers west starting in 1850.

The federal land grants set aside for the constructions of the railroad as shown in Figure 2.1. What is apparent from the map is how the federal government viewed all these lands as being in the public domain, i.e., they were vacant and unoccupied lands. If Native Americans were found living on these lands, they were evicted and had no rights to these lands.

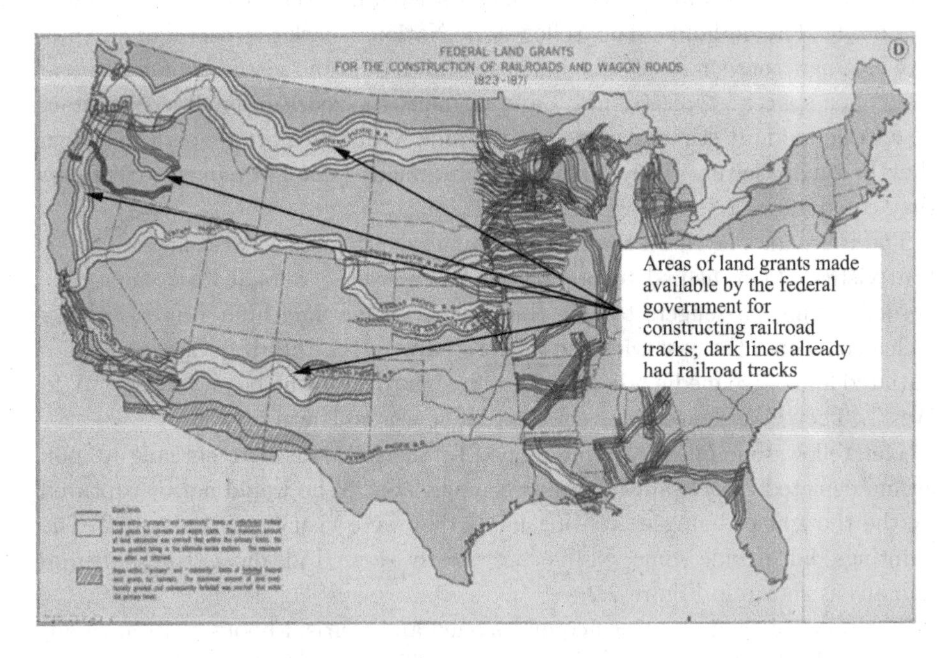


**Fig. 2.1** United States land grants available for the construction of railroads and wagon roads between 1823 and 1871 [31].

To encourage Euro-American settlers to move westward, the images of the "Wild West" had to change. The arts helped with these advertising campaigns. Now the "Wild West" was beautiful and grand—not just wild and untamed. The arts translated these lands for the common man as being beautiful with expansive vistas as far as the eye could see. They suggested that these lands had not been touched by the "hand of man". The Hudson River School, especially Thomas Cole (1801–1848), painted images of the grandeur of the western landscape. The Last of the Mohicans written in 1826 by James Fennimore Cooper delighted several generations of young readers who were to see the wilderness in a new light. Even music was part of our link to a nature without humans. Amy Beach, American composer, wrote in *The Gaelic Symphony* "...I had visions of nature, forests, sometimes vast open spaces, sometimes mountains, always idyllic, organic; I became aware of my soul..." [32]. Those images were of a "sanitized" west since nature did not include native peoples. That same imagery is still used today to attract tourists to visit western U.S.

When the U.S. Congress established the first protected area for recreation in the Americas, i.e., Yellowstone National Park, this grand and pristine view of western landscapes became entrenched in the minds of Euro-Americans and European tourists. In 1872, Yellowstone National Park was created and it was the first reservation of wildlands to be visited for recreation purposes only (see section 2.2.1). The Park was established on federal government set-aside land, i.e., public domain lands. The Yellowstone National Park was especially attractive as a new tourism area. The images of miles of grand vistas, mountains and wide open spaces—typically not found in industrialized Europe—were spectacular to be hold. It stimulated people to think about recreation at a time when they had more leisure time. They were also making enough money working so they could afford to travel westward by train.

The railroads played an important role in advertising the west and its beautiful vistas. They helped to popularize Yellowstone National Park since they needed paying passengers to ride their trains. They had built railroad tracks which connected the east with the western parts of the U.S. Railroad companies initiated a massive media advertising campaign to show the beauty of the "Wild West". They also showed how comfortable it was to travel by train.

The Yellowstone images commissioned by the railroads were enticing to look at and depicted their beautiful and sweeping vistas. Who would not be captured by the thought of living or visiting lands that were visually so heavenly? The paintings did include some wildlife but mostly showed idealic settings with no humans as shown in Figure 2.2.

The railroads had paintings commissioned and tourism books written to advertise and attract people to travel westward. These scenes depicted the west as beautiful and harmless with wide open spaces and "no people". These imageries were a total contrast to the already crowded Europe. Of course, they forgot to

**Fig. 2.2** Images of Yellowstone commissioned by the railroads.
Left: by Thomas Moran, Tower Falls (1872); Right: by Thomas Moran, The Grand Canyon of the Yellowstone (1872). Source: The Anthenaeum. www.the-anthenaeum.org.

paint in the Native Americans and the other wild animals, e.g., buffalo, wolfs, that lived on these lands. Railroads were successful at getting Euro-Americans to become visitors in America instead of being a tourist in Europe. Advertising that focused on the beauty of the west hid the large profits that railroad companies could generate when train ridership increased (//plainshumanities.unl.edu):

> "The promotion of Western scenery by railroads led to the idea of America's National Park system, although the underlying purpose was profit driven, not the preservation of natural resources. Railroads encouraged Americans to forgo the fashionable European Grand Tour and to discover the Western wonders of Yosemite and Yellowstone that they advertised as rivaling the antiquities of the Old World."

### 2.1.3 Becoming Civilized: Redemption and Westward Migration

Not all the images of the western U.S. showed beautiful scenery. Some of the paintings reinforced the idea that it was the duty and responsibility of Euro-American settlers to move out west to civilize these lands, e.g., it would fulfill their Manifest Destiny. Historian Sandweiss [33] from Amherst College wrote how John Gast's 1872 painting, entitled the *Spirit of the Frontier*, was commissioned by George Crofun who wanted to include it in his guidebook.

Gast's painting of the west was included in a travel guide to attract settlers to move west (see Fig. 2.3). The painting depicted a land that would be easy to tame because American Indians and bison would flee from the settlers who had the right, i.e., Manifest Destiny, to possess these lands. It was the duties of Euro-American settlers to civilize this region, i.e., move it from the dark to the light, by bringing the accouterments of technology, e.g., railroads, telegraph lines, and religion. Sandweiss describes the goddess-like figure, with a "the Star of Empire" on her head, as depicting progress [33]:

> "In her right hand she carries a book—common school—the emblem of education and the testimonial of our national enlightenment, while with the left hand she unfolds and stretches the slender wires of the telegraph, that are to flash intelligence throughout the land. The Indians flee from progress, unable to adjust to the shifting tides of history."

**Fig. 2.3** John Gast, *American Progress*, 1872.
Source: https://en.wikipedia.org/wiki/File:American_progress.JPG.

The view that Euro-American settlers had a right and a duty to possess lands in North America, because its current owners were "savage" and "not civilized", permeates the early history of European settlements of the Americas. The ideas behind the Manifest Destiny can even be found in letters written by Jefferson. In 1785, Jefferson wrote [30]:

> "I beleive [sic] the Indian then to be in body and mind equal to the whiteman," Jefferson wrote to the Marquis de Chastellux. Only their environment needed to be changed to make them fully American in

Jefferson's mind. Even though many Native Americans lived in villages and many engaged in agriculture, hunting was often still necessary for subsistence. It was this semi-nomadic way of life that led Jefferson and others to consider Indians as "savages." Jefferson believed that if Native Americans were made to adopt European-style agriculture and live in European-style towns and villages, then they would quickly "progress" from "savagery" to "civilization" and eventually be equal, in his mind, to "white men."

Many writings from this time period support the prevalence of Manifest Destiny in the minds of Euro-American settlers and their leaders:

- "...there was a heroic struggle to subdue the sullen and unyielding forest by the hand of man ... to make it something better than it was ... openings where God could look down and redeem the struggling inhabitants...etc." (W.Penn 1686) [34]
- "...the making of new land demonstrates the direct causal relationship between moral effort, sobriety, frugality, industry and material reward." "..(clearing the land)...tended to measure moral and spiritual progress by progress in converting the wilderness into a paradise of material plenty." (Franklin 1810) [35]
- "What good man would prefer a country covered with forests, and ranged by a few thousand savages to an extensive Republic, studded with cities, towns, and prosperous farms, embellished with all the improvements which art can devise or industry execute ...and filled with all the blessings of liberty, civilization and religion?" (Andrew Jackson 1830) [36]

It is worth taking a brief intermission to review some of the specific actions taken by the Euro-American settlers to fulfill their "Manifest Destiny". These actions had a goal to eliminate Native American culture and traditions, their livelihoods and to take possession of their lands. This goal, as depicted in John Gast's 1872 painting, was under the guise to civilize Indians. Native Americans could not continue to live and survive as they did since it was not civilized. Native Americans living as they had for several hundred years only impeded *Progress* as defined by Euro-American settlers. Being civilized meant that you had to act, think and behave like a European. This meant that you needed to adopt the European view of nature, i.e., nature is isolated from humans and nature needs to be controlled and plucked just like a "garden" for its fruits. You could not retain Indian views of nature and be civilized (see section 5.2).

How the wars on Native American culture and traditions were conducted will be summarized in the next section. These stories cover the American Indian termination, relocation and assimilation policies of the U.S. government. These are important stories to tell because they summarize the extent taken by a government to eliminate a group of people. These actions were carried out under the Manifest Destiny mantle so it was the duty of the government and its citizens to civilize conquered peoples. The Native Americans were the up-river people who had no rights despite all the treaties that they had signed with the U.S. government.

## 2.2 War on Native American Cultures and Traditions

Native American life changed dramatically subsequent to the arrival of the European colonialists. These changes were institutionalized in 1824 when President James Monroe established the Bureau of Indian Affairs "to conduct the nation's business with regard to Indian affairs" [37]. It is noteworthy that when President Monroe established the BIA, the U.S. was at war with American Indians so the BIA was located in the Department of War. The apology made to Native Americans by Kevin Gover (the Assistant Secretary-Indian Affairs, Department of the Interior, Bureau of Indian Affairs) suggests how dramatically the Indian way of life changed [37]:

> "After the devastation of tribal economies and the deliberate creation of tribal dependence on the services provided by this agency, this agency set out to destroy all things Indian...This agency forbade the speaking of Indian languages, prohibited the conduct of traditional religious activities, outlawed traditional government, and made Indian people ashamed of who they were. Worst of all, the Bureau of Indian Affairs committed these acts against the children entrusted to its boarding schools, brutalizing them emotionally, psychologically, physically, and spiritually. The trauma of shame, fear and anger has passed from one generation to the next, and manifests itself in the rampant alcoholism, drug abuse, and domestic violence that plague Indian country. Many of our people live lives of unrelenting tragedy as Indian families suffer the ruin of lives by alcoholism, suicides made of shame and despair, and violent death at the hands of one another. So many of the maladies suffered today in Indian country result from the failures of this agency. Poverty, ignorance, and disease have been the product of this agency's work.... "

Section 2.2.1 of this book will summarize several approaches used by the U.S. government to control lands and resources belonging to hundreds of Native

American tribes. The goal was to ensure that American Indians would not be in a position to take back their lands and resources. This required the elimination of Indian culture and traditions. We will first briefly describe the U.S. termination, relocation and assimilation policies used to eliminate Indian identity and "steal" their lands and resources. This will be followed by stories of practices to eliminate their foods of cultural significance and essential for American Indian survival: killing of the buffalo and then the loss of salmon as dams were built on tribal lands. All of these activities were justified under the colonial "Manifest Destiny" doctrine.

### 2.2.1 U.S. Relocation, Termination and Assimilation Policies

Whenever Native Americans and the U.S. government clashed, it was not good for the Native Americans [38]. The 1492 populations of Native Americans, now comprising the U.S., were estimated to be over 5,000,000 people. Other estimates had as many as 60 million Native Americans living in the present-day U.S. This high original estimate was drastically reduced because Native Americans died faster than the timing of European colonialists visiting other regions of North America. By the time many white people encountered American Tribes farther west, their population density had already been decimated by European-introduced diseases. By 1900, it was estimated that only 250,000 Native Americans were alive in the contiguous U.S. Most American Indians were not immune to the diseases introduced by the European colonialists. In addition to the lack of resistance to European diseases, starvation contributed to the dramatic decreases reported in the size of the American Indian population in the Americas at this time.

There was a major treaty-making period in the mid-1800s, tribes agreed to sell lands for reduced homelands in exchange for promises for economic development, educations, health care, and other items in the treaty-making process. This reduction in land holdings held by Native Americans was considerable. The total hectares of land in the U.S. are 0.9 billion (or 900 million) ha. All of this land was originally held by the Native Americans. By 1881, the Native American land holdings had dwindled to 63 million ha. By 1934, the land holdings diminished to 20 million ha. The amount of land held by Native Americans today is approximately 2% of the total land they originally owned.

Placing American Indians on reservations made it easy for the government to ignore the treaties that they had signed with many tribes. The termination, relocation and assimilation policies implemented by the U.S. government were designed to wrestle control of lands and resources from their original owners for the benefit of the new U.S. citizens. These policies were designed to eliminate any chance that native peoples would rise to demand their lands back. In addition to the termination and relocation policies, the U.S. government made systematic efforts to civilize native peoples so they would become part of the "melting pot" of the new U.S. (see section 10.1). These processes had goals of eliminating native people's cultures and traditions. If you can make the people "part of the melting pot", you do not need any special rules to ensure their rights to their lands, since they would be part of the new country that was emerging in the Americas. The reader is encouraged to read more of the large existing literature that recounts these policies and their impacts on Native Americans. We do not have the space in this book but can only briefly introduce these policies and their impacts in the next section.

#### 2.2.1.1 Relocation and Loss of Customary Lands

Many reasons can be recounted for the relocation of American Indians onto reservations. The European colonialists coveted the lands and resources they found in North America. The Euro-American settlers were especially happy when gold was discovered on these lands. In 1849, gold was discovered at Sutter's Mill, which started a gold rush into California. Shortly after that, new gold discoveries in the Northwest and Alaska caused a rush northward. As this wave of gold miners and pioneers flooded north, much of the land appeared unoccupied to them on any permanent basis. And because of this apparent vacancy, it was very easy for those Euro-American settlers and prospectors to take ownership. They did not have to immediately fight someone to take possession of lands. As it turns out, tribes were seasonally using those lands as important food-gathering areas.

Even treaties between the tribes and the U.S. government did not stop the loss of tribal lands. Tribes were put on reservations that were much smaller in land area relative to what the tribes had been using as their customary lands. In fact even tribes that did not make treaties were put on reservations. The Colville Tribes did not make treaties, primarily since they were amongst the last lands settled in the U.S. due to their isolation and distance from main Euro-Americans' pioneer migration routes. The Colville Tribes also put up resistance and held out behind the natural barriers of the Okanogan and Columbia Rivers in north central Washington. However, this all changed and the Colville Reservation was created in 1872. The Euro-American settlers wanted the lands that the Colville Tribes lived on. For many tribes, the traditional food-gathering sites were prime agricultural lands for the incoming settlers. So these were the first lands to be taken from the reservations because of their agricultural values.

Despite being moved onto reservations, the tribes still did not own these lands and they did not own the resources found on the reservation. The U.S. adopted special property policies and laws for how to control contemporary tribal lands. These policies and laws differ from what is found on normal non-Native lands (see section 5.3.1). Tribal lands are mainly under the control of the U.S. Bureau of Indian Affairs based in Washington DC. The U.S. government signed and ratified treaties with Native Americans to essentially purchase millions of hectares of lands with promises that have, for the most part, never been kept.

These treaties also gave the U.S. government the right to evict Indians from their lands. Between 1777 and 1871 [39]

> "The federal government signed more than 400 Indian treaties ..., with tribes usually receiving various payments and benefits in return for ceding land... State governments, settlers, and businesses pressured the federal government to seize Indian lands for their own use, and more than 100,000 Indians from the Southeast were pushed off of their lands and moved to reservations west of the Mississippi River."

In the 1830s, all tribes were removed from east of the Mississippi River and many were sent to the Oklahoma Territory. In the West, many tribes were relocated to reservations on greatly reduced land areas. The ceded land area compared to the current Yakama Nation reservation is shown in the map (Fig. 2.4). The reservation lands represent approximately 11% of the ceded land area (Fig. 2.4). The gray area demarcates the current Yakama Reservation area and the dark line demarcates the ceded area in Washington State. This is of course

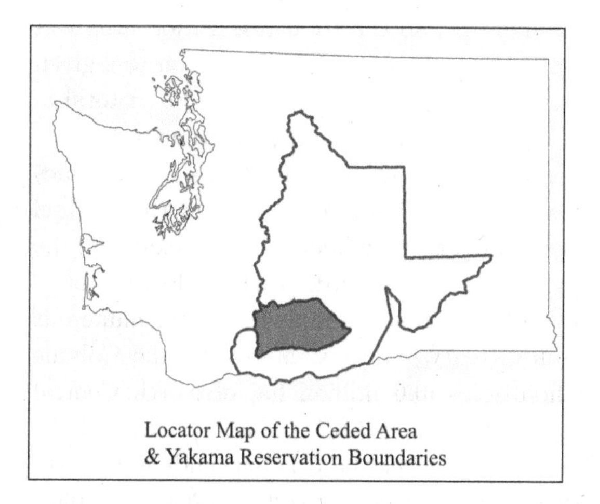


**Fig. 2.4** Yakama Reservation boundaries and ceded lands.
Figure courtesy: John D Tovey (after www.yakamanation-nsn.gov/docs/CededMap0001.pdf.)

not the total land area that had been used to collect resources by tribes now living on the Yakama Nation reservation (i.e., it extended beyond even the red-bounded area). These reservation lands were a small fraction of what each tribe had historically used. For the Yakama Nation, they gave up "over 4.9 million ha of land. But tribal elders have said that their distance of travel sometimes took them as far north as Canada and as far south as California." (www.yakamanation-nsn.gov/history2.php). Therefore, the amount of land area lost by each tribe was much larger than what these numbers indicate.

It was not uncommon for lands designated to become reservation lands to be further reduced in area before the papers were officially signed. Some tribes, like the Yakama Nation, had their reservation lands made available for white settlement (www.yakamanation-nsn.gov/history2.php).

> "Although the Treaty was signed on June 9, 1855 it did not become valid until ratified March 8, 1859 by the U.S. Senate and proclaimed law by the President on April 18, 1859. And just one month after the treaty was signed Governor Stevens through northwest newspapers declared all ceded lands open and available for white settlement."

It was the norm that tribes lost a considerable amount of their customary lands when they signed treaties with the U.S. government.The tribes, e.g., Cayuse, Umatilla, Walla Walla, who form the Confederated Tribes of the Umatilla (CTUIR) ceded 2.6 million ha of land in exchange for 206, 390 ha of reservation land [8]. In fact the surveyors incorrectly measured the reservation boundaries, so these tribes only received about half of this land (99,148 ha). Similarly to what occurred with the Yakama Nation, the CTUIR also had parts of the reservation opened up to non-Indian immigrants, so the total reservation area was further reduced in size. For example, the city of Pendleton, Oregon was given 259 ha of CTUIR land [8]. Eventually the reservation was reduced to a total of 63,940 ha or just a little over 2.4% of the original customary land area.

To re-emphasize again, even when lands were set aside for tribes, many times the total land area of the reservation decreased several times after the original land area had been identified as "designated tribal homeland". For example, the Colville Reservation in "1872 through an Executive Order by President Grant... covered close to three million acres [1.2 million ha], lumping together numerous tribes who were not yet party to any treaty." Today, however, "the Colville Indian Reservation spans 1.4 million acres [0.6 million ha] of North Central Washington primarily in Okanogan and Ferry counties"; the area of the Colville Reservation was reduced only three months after the original lands were dedicated to establish the reservation (www.colvilletribes.com/2011__2012_ceds.php). In this case the driver for this significant reduction in land area of the reservation was the discovery of gold on the Colville Reservation.

A few tribes lived far enough away from the European colonialists so they had less of their lands taken away from them. The Hoopa Valley Tribe lives in present-day California. This tribe has a reservation that is 38,850 ha in size which is about 50% of their customary territory [8]. The Hoopa Valley Tribe has a smaller percent of its lands (3%) "owned in fee simple by non-Indians while the Tribe still owns 95 percent of the reservation in trust with the United States" [8]. They have fewer non-Indians living on their reservation lands because of their remoteness.

Our next story documents how the establishment of the Yellowstone National Park in 1872 resulted in the relocation of Native Americans who had used the park area going back more than 11,000 years [40]. This occurred even though [40]

> "... many groups of Native Americans used the park as their homes, hunting grounds, and transportation routes."

Figure 2.5 is a picture of Washakie and his warriors taken in 1871 in what was to become Yellowstone National Park (Source: [40]). It shows the large number of tribal people who lived and hunted in this newly designated park but were soon to be no longer allowed on their customary lands.

**Fig. 2.5** Photo of Washakie and his warriors in 1871 in Yellowstone National Park [40]. Photographer: William Henry Jackson (1871).

Since the U.S. federal government was at war with the Native Americans, the military was used to enforce who would have access to the newly formed Park territories. The military built a War Department Station at Old Faithful in Yellowstone. This Station was used to house the military so they were well positioned to remove Indians (see Fig. 2.6 left) and to keep Indians from re-entering the park (see Fig. 2.6 right).

**Fig. 2.6** Left: War Department Station at Old Faithful (Haynes; ~1917); Right: Colonel Gardiner—took part in the Nez Perce retreat across Yellowstone National Park (Photographer unknown; ~1877).

Source: NPS 2012.

In 1886, or 14 years after the establishment of the Yellowstone National Park by the U.S. Congress, Yellowstone was placed under the management of the United States Cavalry. The reasons given for this decision were the large-scale poaching of large animals and other administrative needs of Yellowstone National Park.

There is a reason why the hats worn by today's Park Service employees look like the hats worn by the federal troops sent to protect this new park! The federal troops were from the cavalry and wore campaign hats that were then adopted as symbols of authority by the Park Service [41]. They still wear these hats today.

The Cavalry managed and guarded the gates and borders of Yellowstone until 1917. Native Americans were not allowed to enter Yellowstone National Park and to practice their traditional culture until just 11 years ago—more than 115 years after they were originally evicted [42]. This situation did not change until quite recently [42]:

> "In 2001, Yellowstone National Park changed its entrance fee policy to allow members of 'affiliated' tribes to enter the park for traditional purposes without paying the recreation fee".

What happened during the establishment of the Yellowstone National Park, as well as when tribes were evicted from their customary lands, are emblematic of the Euro-American's approach towards indigenous communities. Native Americans were forced off their lands, or had their land area drastically reduced in extent. Tribes had to live on 2.4% to 11% of their total customary land area or 50% if they lived in more remote areas or when fewer non-Indians wanted their lands. They lost all rights to live in the ceded lands. They also were unable to collect traditional foods from their ceded lands even though treaties gave them rights to collect resources from the unclaimed ceded lands. Tribal leaders had reserved the right to fish, hunt and gather their traditional foods in the treaties they signed. The caveat in the treaties was that they were allowed to hunt and gather in unclaimed ceded lands. It didn't take long for nearly all these lands to be claimed and thus little land was left for tribal people to gather their customary resources. Today, these treaties are interpreted to allow tribes to hunt and gather on "public" lands rather than "unclaimed" lands (e.g., Forest Service, National Park Service, Bureau of Land Management, State and local lands).

These stories exemplify our river metaphor where the up-stream people are the Native Americans and the down-stream people are the Euro-American settlers and the U.S. government.

#### 2.2.1.2 Termination

The net effect of the U.S. policies from the 1700s onwards to 1975 was to effectively preclude tribes from developing and to benefit from their own resources. This stopped tribes from developing their own economies. U.S. policies were generally intended to strip away anything valuable that Native Americans might own. They also had goals to move tribal people onto smaller and smaller areas of lands, and lands that at this time had no or little perceived value. It was also customary to place tribes with long standing feuds on the same reservation in the hopes that they may just kill each other off. Finally, even these lands with no value were to be taken from Native Americans when the termination policy of the 1950s began to be implemented.

In the 1950s and 1960s, the federal government terminated the existence of 109 tribes. House Concurrent Resolution 108 states that the government would pursue a policy to end Native American status as wards of the government. These policies remained in place until there was a policy shift in 1975 to end the termination of tribes and to pursue a policy to strengthen tribal governments. This was marked by the passage of the 1975 Indian Self-Determination and Education Assistance Act; so tribes were now in control to begin the process of rebuilding their societies and economies [43].

Even in instances where tribes did retain some assets, the Bureau of Indian Affairs adopted a lease policy which effectively made tribes passive owners of their lands and resources. Tribes collected relatively small rents and royalties, with the bulk of benefits and profits and jobs going to non-Indians. Furthermore, technology was kept in the hands of outside corporations and tribes were kept in the dark.

The impacts of evicting Native Americans from their lands were acknowledged in 2000 by Glover when he apologized to Native Americans for the actions of the Office of Indian Affairs (BIA). He spoke and wrote [37]:

> "...From the very beginning, the Office of Indian Affairs was an instrument by which the United States enforced its ambition against the Indian nations and Indian people who stood in its path. And so, the first mission of this institution was to execute the removal of the southeastern tribal nations. By threat, deceit, and force, these great tribal nations were made to march 1,000 miles to the west, leaving thousands of their old, their young and their infirm in hasty graves along the Trail of Tears..."

After removing tribes from their homelands and onto reservations, and destroying their traditional economies and ways of existence, the tribes were left in deplorable conditions. The assimilation of Native Americans into Euro-American societies, however, was just beginning. Children were especially vulnerable to these assimilation policies. To make Indians into a good model of Euro-Americans, they needed to be removed from their families and taught a useful trade. They were punished when they spoke their native languages and were not allowed to practice their cultural traditions.

Some tribal people believe that pragmatic decisions made in BIA run schools contributed to the loss of tribal languages. Teachers at BIA schools were teaching students from 15-30 different reservations. The teacher needed a common language to speak to tribes who had many different languages. My grandmother (John D Tovey) and stories from other families indicate that for the most part students were allowed to speak their own languages but when they were not in class. But then again it was difficult to speak to your friends if they spoke a different language, so English became the "lingua franca" out of necessity.

To civilize Native American children, any remnants of culture, languages and traditions were to be removed. Civilized meant you looked, acted and behaved like a Euro-American. What is amazing is that these policies continued until the last decade. They were not just something that happened several hundred years ago. A brief review of the assimilation policies is warranted to show that the assimilation policies targeted the elimination of culture which is a critical element for the survival of a people.

#### 2.2.1.3 Assimilation

> *"Let me be a free man, free to travel, free to stop, free to work, free to trade where I choose, free to choose my own teachers, free to follow the religion of my fathers, free to talk, think and act for myself—and I will obey every law or submit to the penalty"*

~ Quote from the 1879 Chief Joseph's speech, Lincoln Hall, Washington DC ~

At the time that tribes were being relocated and terminated, the U.S. adopted policies intended to mainstream Indians into the U.S. society. This did not mean that Native Americans were to become U.S. citizens once they were "civilized". Tribes were not given U.S. citizenship until the 1924 Indian Citizenship Act but only if they were born after its effective date (one could also earn citizenship by enrolling in the military, assimilating, etc.). It wasn't until 1940 (Nationality Act 1940) that all Indians were given citizenship as long as they were born within the contiguous U.S. The story of assimilation of Native Americans is not a pleasant story but has to be told.

Assimilation policies were designed to complete the Native American transformation or breaking away from past traditions and forcing them into new traditions. Several strategies were attempted to forcibly assimilate Native Americans into the Euro-American society and lifestyles. This assimilation was the ultimate form of becoming civilized. The primary mission was to abolish all traces of traditional tribal cultures and to physically relocate Native Americans into urban areas. The hope was that Indians would be absorbed into mainstream U.S. society and disappear - the Indian problem would then be solved. Many Indian children were virtually kidnapped and placed into foster homes and adopted out to non-Indian families.

Manifest Destiny also encouraged the conversion of Native Americans' religions to European-derived ones. It was used to justify eliminating Native Americans and any trace of their culture as noted in the 1707 writing of John Archdale [44]:

> "And courteous Readers, I shall give you some farther Eminent Remark hereupon, and especially in the first Settlement of Carolina, where the Hand of God was eminently seen in thinning the Indians, to make room for the English... it at other times pleased Almighty God to send unusual Sicknesses amongst them, as the Smallpox, etc., to lessen their Numbers..."

The churches were an important tool used by the federal government to control Native Americans. They believed that tribal religious and cultural beliefs were retarding tribal progress. To solve this problem, churches were given operating franchises to tribes so they could convert Indians. This was a federally regulated process so churches had to get a federal permit to proselytize Indians.

The "Manifest Destiny" was now being implemented by religious institutions in charge of converting Indians to western European beliefs. Traditional Native American religions were considered to be pagan and unacceptable by Euro-Americans. They did not recognize that the Native American culture has a diverse variety of nature-based religions among the different tribes (see Box 4).

**Box 4: Native Spirituality Today**

"Historically, those who weren't American Indians were confused, sometimes downright antagonistic, about the spiritual practices of American Indians. They had a difficult time understanding the spiritual practices of native people that were so different from the religious practices they knew and practiced. They were used to attending church and studying the scripture—the Bible or the Torah. With American Indian spiritual practices, nothing was written down and worship wasn't necessarily held in buildings. Oftentimes these practices were based in nature, held outside and, then as now, information was imparted on a need-to-know basis.

...It wasn't until 1976 when the American Indian Religious Freedom Act (AIRFA) was passed that native people were guaranteed the freedom to practice their traditional religions. ...

...Washington State's First People engage in their own unique spiritual and religious practices. Some practices are ancient, like the Washat longhouse religion that flourishes today among the Confederated Tribes and Bands of Yakama Nation and Upper Plateau peoples. ...Some spiritual practices are newer and are a hybrid of traditional beliefs coupled with mainstream religions. The Shaker Church is an example of this..."

(from Kimberly Craven [45], Sisseton-Wahpeton Oyate)

The Euro-Americans did not appreciate the natural or environmental aspects of Native American religions. This is understandable since most of the European religions were formed at a time when the environment and nature were not an important aspect of religion, e.g., they are not part of the scriptures. Even religions that formed after the European conquest of other global societies acknowledged environmental problems but mostly did not formally include it in their religious practices.

A large percentage of Native Americans today practice their traditional religion and the adopted religions that were introduced to them by the Euro-American settlers. Tribes practice several religions at the same time and accept parts of another religion without eradicating their own religion (see Box 4). Several New Age religious practices (of both Native Americans and other people) also encompass various aspects of Native American religions.

The assimilation policy had a goal of converting the Native Americans to think, act, look and behave like western Europeans (see Fig. 2.7). In the boarding schools, children were made to look like and act like people who you could have been found walking the streets of Europe. When dressed in Western Europe clothing, anyone could appear civilized.

**Fig. 2.7** Left: Chiricahua Apaches four months after arriving at Carlisle; Right: American Indians dressed in western clothing.

Source: Richard Henry Pratt Papers, Beinecke Rare Book & Manuscript Library, Yale University; //en.wikipedia.org/wiki/File:Assmilation_of_Native_Americans.jpg.

The idea behind most of this was to remove the cultural and traditional underpinnings of Native American children, i.e., civilize them by having them adopt the lifestyle and religions of their conquerors. Indian children were virtually kidnapped and sent off to far away U.S. boarding schools to further erase all tribal culture and to cut off ties with their families. Others were sent off for adoption into non-Indian families. It was against the law to speak your own language, to wear traditional clothes, to wear your hair in a traditional fashion, or to even have gatherings of more than three Indian males in one spot. As with the Blacks in the U.S., Indians were prohibited and segregated and were often banned from eating in restaurants and cafes and bars. One of our authors (John D Tovey) recounted "My Uncle Williams, while in full U.S. military uniform, was attacked while using a pay phone in Western Florida in the 1940s because he was on the wrong side of the road".

Their children were removed and educated in western educational institutions or boarding schools that were run by different western religious denominations. Boarding schools were created to separate children from their families and culture to assimilate them into the mainstream society. Thousands of Native children were forced into attending these schools and did not return to "their homes until they were young adults" [46]. Boarding schools were established all over the country [46]

"By 1909, there were over 25 off-reservation boarding schools, 157 on-reservation boarding schools, and 307 day schools in operation."

These boarding schools were first established by Christian missionaries who were paid by the federal government to educate Native American children who lived on reservations. Eventually the BIA founded more boarding schools off-reservation, similar to the Carlisle Indian Industrial School (see Fig. 2.8). The photo shown below is from the Carlisle Indian Industrial School in Pennsylvania that was one of the first off-reservation boarding schools founded in 1879 [46].

**Fig. 2.8** Carlisle Indian Industrial School in Pennsylvania (c. 1884).
Source: Anne Ely photograph Album Page 21, RG851s, US Army Heritage & Education Center. http://cdm16635.contentdm.oclc.org/cdm/compoundobject/collection/p16635coll15/id/1002.

Church groups were given franchises to convert Native Americans into Christians. This quote by Adams published in 1995 summarizes the process of civilizing the Indians in boarding schools [47]:

"Indians must be taught the knowledge, values, mores and habits of Christian civilization... Since the days of the common school movement, the schoolhouse had come to achieve almost mythological status. Reformers viewed it as a seedbed of republican virtues and democratic freedoms, a promulgator of individual opportunity and national prosperity, and an instrument for social progress and harmony. Moreover, because of the common schools alleged ability to assimilate, it was looked upon as an ideal instrument for absorbing those peoples and ideologies that stood in the path of the republic's millennial destiny."

Studies done in the 1970s showed that up to 25%–35% of all Indian children had been removed from their families and placed into non-Indian care [48]. What is amazing is that the number of Native Americans enrolled in these schools only peaked in 1970. In fact, some American Indian children were still in boarding schools in 2007!

The intent of the boarding schools for the young Indian females was to train them to be future wives for non-Indian settlers. The government even sponsored posters advertising their availability. Marrying Indian females to white men had significant impact on how the population of Native Americans decreased. Canadian enrollment laws for Indians used to only allow Indian enrollment to follow patrilineal lines. If an Indian woman married a white man she lost her "Indianness". But if a white woman married an Indian man she became an Indian. This proved to reduce the numbers of Native American people drastically because children of women married to white men were no longer considered Indian, and white women didn't want to suffer social stigmas. This situation also was difficult to deal with culturally since many coastal tribes are actually matrilineal societies.

A United Nations' report summarized the history of the boarding schools and the attempts to assimilate Native Americans into the western European culture in 2010 [46]:

- "... 19th and early 20th centuries, Native American children were forcibly abducted from their homes to attend Christian and government-run boarding schools as state policy. The boarding school system became more formalized under the Grants' Peace Policy of 1869–1870, which turned over the administration of Indian reservations to Christian denominations. Funds were set aside to erect school facilities to be administered by churches and missionary societies.
- ... The rationale for off-reservation boarding schools was 'Kill the Indian in order to save the Man' as well as 'Transfer the savage-born infant to the surroundings of civilization, and he will grow to possess a civilized language and habit.' The strategy was to separate children from their parents, inculcate Christianity and white cultural values upon them, and encourage or force them to assimilate into the dominant society. For the most part, schools primarily prepared Native boys for manual labor or farming and Native girls for domestic work. Children were also involuntarily leased out to white homes as menial labor during the summer rather than sent back to their homes.
- ... Boarding schools were administered as inexpensively as possible. Children were given inadequate food and medical care, and conditions were overcrowded. According to the Boarding School Healing Project (BSHP) Native children in South Dakota schools were rarely fed and as a result, children routinely died in mass numbers of starvation and disease. Other children died from common medical ailments because of medical neglect.

In addition, children were often forced to do grueling work in order to raise monies for the schools and salaries for the teachers and administrators. Children were never compensated for their labor.

- ...Many survivors report being sexually abused by multiple perpetrators in these schools. However, boarding school officials refused to investigate, even when teachers were publicly accused by their students. There are reports that both male and female school personnel routinely abused Native children, sometimes leading to suicides among these children."

Many of these dislocated and shattered tribal communities today really have not yet recovered from these actions. There is still massive unemployment on many of these reservations and there has never been a replacement economy created to take the place of the buffalo-based economy which existed prior to the arrival of the European colonialists. The buffalo-based economy was central to many Native American tribes which explain why such efforts were made to remove buffalo from the Great Plains. This will be recounted next.

### 2.2.2 Removal of Buffalo for "Manifest Destiny"

One of the first tribal resources to be eliminated was part of a military strategy to remove buffalo from the Great Plains in the 1800s. The buffalo, a resource that favored one group, the Native Americans, was eliminated and replaced with farms and cattle to favor another incoming group, the settlers. This pattern would be repeated in the Northwest with salmon which were sacrificed so that dams could be built to produce electricity. The following information for buffalo was taken from a fact sheet produced by the South Dakota Department of Game, Fish, and Parks [49]. For example, on the Great Plains, buffalo were once numerous and were central to the lives of Native American people. Bison populations are estimated to have numbered 60 million, but by the late 1800s this number had dwindled down to only 1,100 animals.

Over thousands of years, the Native American culture and society on the Great Plains evolved to maximize utilization of the vast herds of buffalo. The Native American villages were mobile; they could follow the herds and a nomadic culture evolved to take advantage of the mobile bison herds. The buffalo provided the food, clothing, housing, tools, and weapons needed for their survival. Their society was organized around the buffalo; it was a part of their religion and core values as a society. They hunted communally and the buffalo was shared with the weak and strong, and with the old and the young. This societal reciprocity and high value placed on sharing was highly valued. All aspects of their life depended greatly on the buffalo.

My own people (Michael Marchand), now comprising the Colville Tribes of eastern Washington, also hunted buffalo each year. About half the tribe fished for salmon while the other half rode by horses onto the Great Plains to hunt buffalo and to trade. The buffalo had a major impact on the lives of many Native Americans.

To defeat the Plains tribes, the U.S. had a policy to exterminate the buffalo in order to take away one of the key resources of the Native Americans. Buffalo were hunted and slaughtered to near extinction. Commercial hunters were encouraged. Indians were also involved in this carnage and also participated in the commercial operations. This elimination of the buffalo had two benefits for the U.S. federal government. One was to defeat the Native Americans and help force them onto reservations, the other was to open up these grasslands for domesticated cattle and for agriculture.

Likewise it was also common for the U.S. military to slaughter the horses of Native Americans to take away their mobility. Horses were central to their economies, to gather foods, to hunt, and to trade.

Being faced with starvation and freezing weather, many tribes were forced onto reservations and were made dependent on U.S. food and clothing handouts. Many of their societies were on the brink of collapse. Removal to reservations had cleared the way for settlement of the West, but, in spite of the move to the reservations, the displaced Native Americans were still in dire need.

Not only buffalo but salmon was lost from the diet and culture of Native Americans living in New England, Atlantic Maritime Provinces of Canada and in the Pacific Northwest (PNW) U.S. In contrast to buffalo which were systematically exterminated, the loss of salmon was not a targeted strategy to destroy Native American cultural foods. Even back in the 1700s on the east coast of North America, over-fishing of salmon by Euro-Americans to supply British fish markets caused the collapse of the salmon fisheries [15].

On the west coast of the U.S., building dams to provide inexpensive electrical power to the growing urban population centers caused the loss of the salmon fisheries (see section 2.2.4). The pre-European robustness of the salmon fisheries in the Pacific Northwest U.S. is important in understanding the consequences of their decline and will be discussed in the next section. That will then be followed by a brief discussion of the impacts of building the Grand Coulee Dam on tribal lands and tribal loss of cultural resources, e.g., salmon.

### 2.2.3 Removal of Salmon in the Pacific Northwest

The story for salmon based tribes in the Pacific Northwest (PNW) U.S. is very similar to the story of the buffalo for the Plains tribes and the story of oil drilling on Indian lands. The motivations may have been somewhat different in the details but at the larger framework, the story is very much the same. Tribes had been utilizing a natural resource, the salmon, for thousands of years. However, the incoming settlers wanted to redevelop the natural environment and shift these benefits onto the non-Indian settlers and their businesses and other interests.

Before the construction of dams, armoring, dikes and other barriers, all the rivers and streams had salmon. Many coastal tribes fished for salmon in the ocean. Historically, PNW US tribes have fished for salmon on the Columbia River for at least 10,000 years, since the last ice age at a minimum.

The PNW Native American societies were salmon dependant (see Box 5 and Fig. 2.9). The Columbia River was the backbone for the salmon and the Native Americans. It provided food. It provided the transportation on its waters. The river was central to their existence. The importance of the salmon fisheries is described for Kettle Falls in Box 5. The large size of each salmon was even noted on the Atlantic coast of the Americas. Trefts [15] recounted how the early colonialists to Massachusetts noted in 1634 "... no country known to Europeans yielded more variety of fish, winter and summer ... for trade into other countries."

**Box 5: Great Salmon Fisheries and Kettle Falls**

Kettle Falls was one of the great fisheries of the world. The thundering falls was 15.2 meters tall. The keepers of the falls were the Arrow Lakes people, the Sinixt people. This area was primarily the homeland of the Sin Aikst or Colville Tribe or Band. They also had a salmon chief and he regulated the fishery. They utilized all types of fishing methods, but they were unique in that they utilized basket traps. They made huge baskets 9.1 meters long and placed these at the foot of the falls. When the thousand of salmon made their leaps to get over the falls, thousands of them would fall shot and land in the baskets. Early explorers in the region were astounded at the multitudes of fish, they counted two thousand fish being pulled out of a single trap in one day, and they were also astounded that the fishermen exercised restraint and stopped fishing when the salmon chief so ordered them to do so. They remarked that they did not think that the white man could restrain himself enough to stop. There is a written account from Jesuit Priest DeSmet in 1845, where he wrote [51],

> "My presence among the Indians did not interrupt their fine and abundant fishery. An enormous basket was fastened to a projecting rock, and the finest fish of the Columbia, as if by fascination, cast themselves by dozens into the snare. Seven or eight times a

(Continued)

> day, these baskets were examined, and each time were found to contain about two hundred and fifty salmon..." [51]
>
> Kettle Falls was a major gathering place. Each summer about five thousand Indians would gather there to fish and to trade. It was a major trade center. Tribes visited from all over to see the falls and to trade. Artifacts have been found there from all corners of the continent. Raw materials and products were brought in, there were craftsman at the falls, and trade took place. The salmon themselves were a major trade commodity. Thousands of salmon were caught and most of the fish were dried on racks in the sun and they were put into bales. These bales were stacked up and eventually they would be brought downriver and traded up and down the Pacific coast and also taken into the Great Plains for more trading.

**Fig. 2.9** Salmon Chief Tommy Thomson; Note the size of the salmon hanging from the rafters.

Courtesy: Yakama Valley Regional Library, Accession number 2002-851-296. www. yakima-memory.org.

The 10,000 years of salmon fishing at Kettle Falls ended in 1942 with the construction of the Grand Coulee Dam. Figure 2.10 shows a salmon swimming up the water falls that existed at this location before the building of the dam.

Figure 2.11 is of the same area shown in photo above but after the Grand Coulee Dam was built.

The Dam is the most massive construction project of its time and it is still the largest hydroelectric dam in North America. The dam was built under the

**Fig. 2.10** Former area of the Grand Coulee Dam before the dam was built.
Photo by Frank Palmer, Courtesy Northwest Museum of Art & Culture. Neg L94-40.53.

**Fig. 2.11** Grand Coulee Dam.
Photo courtesy: Michael Marchand.

U.S. Bureau of Reclamation. It is the key for a massive irrigation project that diverts water from the Columbia River southward into the Columbia Basin to support agriculture over a hundred miles from the dam itself.

The WCD Report states, "Due to neglect and a lack of capacity to secure justice because of structural inequities, cultural dissonances, discrimination, and economic and political marginalization, indigenous and tribal peoples have suffered disproportionately from the negative impacts of large dams, while often being excluded from sharing in the benefits" [50]. The U.S. political and legal system had forcibly put tribes under the trusteeship of the U.S. government. Tribes were placed under paternalistic policies which severely limited their ability to defend their own interests to protect their resources and ways of life. The U.S. wanted hydro-power development and the First Nations people were in the way of progress.

In the PNW U.S., the population of Native Americans was estimated to be over 1,000,000 prior to the coming of the white man. By 1850, this number had dwindled to 10,000 Native Americans.

Then there was a major influx on settlers and immigrants into the region and by 2000, the northwest population had grown to 11,000,000 people. Much of this can be attributed to the availability of abundant electricity and the rapidly growing economy made possible by the energy produced inexpensively by hydro-electric dams.

For the Colville Tribes, the loss of salmon was the worst impact of building the Grand Coulee Dam. The dam was built and there are no fish ladders on the structure. The salmon literally swam upstream, ran into the dam and had no place to go in 1942.

So 10,000 years of life based on the return of the salmon came to an abrupt halt that year. The salmon-based culture and economy was devastated. One day they were fishermen, but now there were no fish. The main food source was gone. Salmon was also a trade commodity; the basis for their trade was gone.

For thousands of years the Colville peoples were wealthy and the salmon were abundant, but then they were gone. Their economy is shattered and still has not been replaced.

### 2.2.4 Building Dams on Tribal Lands

Early American cities of the 1700s were often located on rivers. One reason for this was to utilize water wheels to provide power to grind flour. River energy also provides horsepower needed to power early efforts at industrialization, e.g., to produce textiles. This is where the dams were built since the power of the rivers was needed to provide the horsepower for all these industries.

The early use of water wheel technology to power industries located along rivers eventually gave way to hydro-electric power. Water was converted into electricity. The World Commission on Dams Report states that, "The first use of dams for hydropower generation was around 1890." The Report continues, "By 1949, about 5,000 large dams had been constructed worldwide... by the end of the 20th century there were over 45,000 large dams in over 140 countries [50].

This takes us back to the building of the Grand Coulee Dam on tribal lands in the Pacific Northwest U.S. There were many benefits from the building of this dam for the Euro-American settlers who settled the Pacific Northwestern U.S. This dam was considered to be the Eighth Wonder of the World. Its construction actually began in 1933 and final construction of the Third Powerhouse was not completed until 1974. The initial dam construction was completed by 1942. It is still the largest concrete structure in the world and contains 9.2 million $m^3$ of concrete (www.usbr.gov/pn/grandcoulee/pubs/factsheet.pdf). It is 1,178 m wide and 167 m tall. There is enough concrete to build an 18 m wide highway from Los Angeles to New York City (www.usbr.gov/pn/grandcoulee/pubs/factsheet.pdf). It is four times larger than the Great Pyramid. Its total generating capacity is 6,809 megawatts and it is the fourth largest power producer in the world today (www.usbr.gov/pn/grandcoulee/pubs/factsheet.pdf).

The Grand Coulee Dam destroyed the Native Americans way of life. But it allows the non-Indian Northwest economy to flourish and prosper. U.S. Senator Clarence Dill summarized the benefits of the Grand Coulee Dam when he spoke at the Ceremony of Tears in 1940 (see Box 6 and Fig. 2.12).

Many benefits accrued to Euro-Americans living in Washington State from the building of the Grand Coulee Dam. First, people were employed to build the Grand Coulee Dam. The construction work force was 10,000 workers. This dam

**Box 6: Ceremony of Tears**

One of the speakers was U.S. Senator Clarence Dill and the War was on his mind that day, he said, "We can build more airplanes and tanks and can train more pilots for national defense than any other nation of combination of nations, and the quicker we do it the better... We know now that the only thing in this world which Hitler will respect is more force then he controls." [50] Dill acknowledges the terrible loss to the tribes but he hoped that the Indians would realize some benefit from the power that the dam would produce [51]. As a Senator, he could have done more than hope of course, he could have written these benefits into law if he had any real desire and this was never done of course (see Fig. 2.12). That would never happen for another half century of struggle.

High voltage transmission lines went south from the Grand Coulee Dam, crossed the eastern Washington plateau and fed power into the cyclotrons at the Hanford Project. It played a major role in the development of the atom bomb and this led the way into the modern era of quantum physics with the high technology that we take for granted today. Other power went into the newly created aluminum industry which depended upon the abundant hydro-power. This aluminum in turn went to the Boeing aircraft factories of Seattle and elsewhere to build up the war machine. One could logically argue that the present high technology industries of Seattle and Portland can trace their roots back to the construction of the Grand Coulee Dam and also perhaps to the 1940 Ceremony of Tears. One group was supposed to fade away into oblivion and a new people were to emerge out of their sacrifice. From an energy perspective, the power was transferred from the Native Americans and the salmon to the newly emerging economies.

**Fig. 2.12** Colville women at the Ceremony of Tears, 1939.
Courtesy: Northwest Museum of Arts & Culture (Image No. L96-90.38).

provided irrigation for 1400 farms; total irrigated lands amounted to 276,000 hectares, and farm revenues generated $637 Million dollars in income [50].

Subsequent to the construction of the Grand Coulee Dam, seven more dams have been built on the upper Columbia River, the original homeland of the Colville Tribes (see Tab. 2.1). Three of these dams are located in Canada, the homelands of the Arrow Lakes people, now members of the Colville Confederated Tribes.

**Tab. 2.1** List of dams built on the upper Columbia River and the amount of electricity produced by each dam (www.usbr.gov/).

| Name of Dam | Amount of Electricity Produced by the Dam/MW |
|---|---|
| Mica Dam | 1,805 |
| Revelstoke Dam | 1,980 |
| Keenlyside Dam | 185 |
| Grand Coulee Dam | 6,809 |
| Chief Joseph Dam | 2,620 |
| Wells Dam | 840 |
| Rocky Reach Dam | 1,287 |
| Rock Island Dam | 660 |
| Total | 16,186 |

The Arrow Lakes people have been declared extinct by Canada and their lands were confiscated. Currently there is an aboriginal title case pending in the Canadian legal system. The aboriginal, traditional homelands of the Colville Tribes now support a tremendous hydro-power resource, which benefits the entire Pacific Northwest economy today.

The primary benefits of the dam itself targeted off reservation communities. This dam provides many benefits to regional non-Indians. It is able to control regional water supplies and therefore energy production derived from hydro-power. Massive irrigation systems flow south into the Columbia Basin Project to irrigate vast farm lands. Transmission lines carry the power into the Bonneville Power Administration grid system to service the west coast.

There were many negative impacts of building the Grand Coulee Dam on the American Indian tribes who lost their customary lands and an important cultural resource, the salmon. This is another example of the up-river and down-river analogy with tribes being the up-river people who paid all the costs and derived few if any benefits from building a dam on their lands. The negative impacts of building the Grand Coulee Dam on lands of the Colville Confederated Tribes will be described next.

#### 2.2.4.1 Negative Impacts of Grand Coulee Dam

The completion of the Grand Coulee Dam in 1942 represents a major renewable energy project that had major national impact on the U.S. There is a general belief that renewable energy is good, a win-win for all. Hydropower is generally viewed as a clean source of energy. Most of the public is probably not aware of the negative consequences of large hydropower. Hydropower favors some groups and creates problems for others. The Grand Coulee Dam, "provides a vivid example, Native Americans were physically displaced by a project that provided power to industry and households to a city 250 km away and the water and land that had previously supported their livelihoods (particularly for fishing) was dammed and diverted to provide white settlers with irrigated farmland" [50].

For many years, despite being in the shadows of the largest dam in the world, many of the Native Americans did not even have electric service in their homes. The WCD Report states, "the Grand Coulee Dam Case Study reports that the loss of salmon also had severe cultural and spiritual consequences integral to the First Nations way of life" [50].

This was not a new story, and just as it started with the removal of the buffalo, it would be repeated with additional dams and new projects in the future. From 1940 onward, the Native Americans impacted by the Grand Coulee Dam began their long struggle to cope with the new world dropped on them and to make sense of it all again.

Salmon once sustained the Native Americans and their economies for ten thousand years. This all came to a halt in 1941, when the Grand Coulee Dam neared completion. This dam did not have any provisions for salmon passage at all. Thousands of salmon and steelhead showed up to the headwall of the dam and had nowhere to go. This loss of fishery stock followed a similar pattern to what had already happened on the Atlantic coast but back in the 1800s (see Box 7).

**Box 7: Atlantic Fisheries**

When the European explorers first arrived on the Atlantic coast of America, the transformation of river systems by building dams for power triggered the demise of the Atlantic fisheries. Impassable dams were already being built in early 1600s to power mills built along rivers. After 1790, this region began to industrialize and number of dams built increased dramatically. Industrialization meant building dams "throughout New England to supply power to hundreds of cotton and weaving factories and to factories producing firearms, furniture, clocks, machine tools, shoes, and paper".

Trefts wrote how "... routine dumping of refuse, untreated waste, animal carcasses, dyes, chemicals and many other industrial byproducts turned the Merrimack and many of its tributaries into virtual sewers." Fishery stocks crashed or became depleted in the 1800s in the rivers located on the Atlantic coast of the Americas.

The depletion of Atlantic fisheries is in sharp contrast to what the European colonialists found when they first arrived on the Atlantic coast of America. The fish of the Atlantic coast of the Americas was valuable enough that "European explorers... recoup the trans-Atlantic shipping costs." Fish stocks were considered to be inexhaustible even up to World War II and yet fish declines were already recorded in the 1800s.

(from Trefts [15])

One of the greatest salmon fisheries on the planet, the great Kettle Falls fishery, was inundated and there were no more salmon at all. This was all a result of the building of the Grand Coulee Dam. The Native American Arrow Lakes people, the keepers of the falls for 10,000 years, were put into instant poverty, their jobs as fishermen were now gone and their main food source, the salmon, was gone, and their main trade commodity, the salmon, was also gone forever. The great dam forever displaced Arrow Lakes people from their ages old ways of life.

Many communities were also displaced and either disappeared or elsewhere moved to higher ground. Many individual homes and many traditional village sites along the Columbia were also submerged along with associated cultural resources when the dam was built. Cemeteries also had to be moved out of the flood zone. The banks keep slumping in and more ancient remains were uncovered each year.

The fabric of their society was obliterated, everything that was a foundation for their traditions and culture and economies were instantly gone. The results have been devastating. There was massive unemployment and poverty. Their society was made dysfunctional. Social problems were out of control, families broke down, alcoholism was killing people like flies, suicides were high, many went to jail, and it is a sad litany of one bad thing after another.

Many of the young men joined the military. They came from a warrior culture and after all, what else was there to do? Native Americans on a per capita basis serve their country in the military more so than any other group in the U.S. and this continues today.

So, for WW II, the young Native Americans went off to fight in the big war and the dam produced energy for the war effort. The Grand Coulee Dam played a key role in the war effort since it provided the energy for plutonium production and also for aluminum production, both are keys to the victory of the Allied military. But when the young soldiers came home, many got off the troops ships and hitchhiked cross country to get home, they found that their once Great Kettle Falls was no longer there. Kettle Falls was now out of sight and submerged underneath Lake Roosevelt created by Grand Coulee Dam. There were no fish. There was no economy and no one was offering help.

Today, the Grand Coulee Dam can be considered as one of the older renewable projects. Efforts by the affected Native American government to win compensation for damages from the great dam began after the tribe filed an appeal of a government payment to the tribe in 1946.

The issue of payment for lands flooded was not resolved until half a century later, when the US Congress finally passed an act to provide a negotiated compensation amount. The WDC case studies "show that the direct adverse impacts of dams have fallen disproportionately on rural dwellers, subsistence farmers, indigenous peoples, ethnic minorities, and women" [50].

The salmon was gone and 10,000 years of prosperity from the greatest fishery on earth vanished. There were no salmon to eat. The once proud fishermen now had no fish, they were out of jobs. The fish-based economy was in shambles. The entire foundation of the tribe's culture and traditions was shaken to its roots. The reservoir behind the world's largest concrete structure was like a giant moat, it cut off all transportation from the bulk of the Colville Reservation, making economic development a difficult task that has yet to be solved. Two ferry boats now serve the Lake Roosevelt, but they are small and designed for limited car traffic. They are not suited for large volume commercial traffic. The prime river bottomlands with their rich soils were now at the bottom of the lake. Entire communities were relocated and many just disappeared.

The Colville Tribe has been a victim of energy development since 1942, with the construction of the Grand Coulee Dam on its border. This dam created major benefits for the Pacific Northwest in general, but the brunt of the costs fell onto the Colville Tribe [52]. This massive Bureau of Reclamation project had giant positive impacts, creating huge amounts of electricity and providing water from the Columbia Basin irrigated agriculture. It has been a catalyst for development of northwest economies. Unfortunately, the dam stopped all salmon runs, thus the Colville Tribe's entire way of life was destroyed, the fishing-based economy was obliterated, and the tribe's culture and religion, based on the salmon, were also devastated.

The reservoir behind Grand Coulee Dam—Lake Roosevelt—covers 33,306 hectares of land and is 240 km wide (www.usbr.gov/pn/grandcoulee/pubs/factsheet.pdf). It forms an effective moat around the majority of the Colville Reservation. It creates a major transportation barrier which impedes present and future access to the reservation, and hence, tribal economic development.

With the building of the Grand Coulee Dam, losses to the local tribes include not only real estate with mineral rights (gold and other low grade ores), but also the salmon and other native animals used by the tribes for their livelihoods, or for cultural reasons. There is a legacy's of chemical misuse which has created a Super Fund Site in Colville. At the same time, tribes living on the Colville Reservation have to survive on a smaller land base and be resilient (culturally, etc.) from what remains of their customary lands.

This continues today with the addition of wind-turbine power around the Grand Coulee area. A problem occurs in approximately 60–90 days out of the year when there is too much electricity generated by the two separate sources; during 2012 the power company was not going to accept electricity generated by wind power. This additional energy generated by wind cannot be stored, so the generation of electricity at the dam site was halted. This creates a situation where users are actually subsidizing the wind power developers.

The lands above the Grand Coulee Dam are amongst the poorest in the U.S. There is chronic high unemployment and poverty for both Native Americans and the non-Indians living upstream from the dam. Major socioeconomic problems abound, ranging from high suicide rates, drug and substance abuse problems, high dropout rates from school, and the list goes on and on. Billions of dollars of benefits have been streaming out of the Grand Coulee Dam since 1942, but there has been only minimal investment to mitigate the negative impacts to the Colville Tribe and the original salmon-based economies have not been replaced by adequate alternative economies.

There was no plan or intention of replacing the livelihoods of the Native Americans when their economies were destroyed and the region is still economically depressed. The WCD Report concludes that replacing the livelihoods of displaced indigenous peoples requires planning and preparation beforehand, "Regaining lost livelihoods requires adequate lead time and preparation and therefore people must be fully compensated before relocation from their land, house, or livelihood base" [50]. But this was not considered by the U.S. in its hydro- power development policies and the Colville Tribes will likely still be striving for mitigation well into the next century for damages from the Grand Coulee Dam project.

Other negative impacts of building the Grand Coulee Dam on Native Americans are:

- It was not only the loss of salmon and the loss of a major trade center, but also the loss of anadromous fish habitat.
- There were impacts of the upstream and downstream ecosystems and other animals that depended upon anadromous fish populations.
- Decrease of surface water quantity and quality consumed for agriculture and the over-permitting of water resources.
- Continual river bank destabilization has impacted developments, ecosystems, disturbed cultural resources and created the need for river bank stabilization projects including armoring.
- Introduction of widespread intensive agriculture introduced toxic pesticides and fertilization.
- Widespread agriculture also decimated thousands of acres of shrub steppe lands where Native people harvested traditional vegetable root crops, medicines, tools, etc. Agriculture has also uncovered many gravesites and other cultural resources that were buried by the building of the dam.
- The benefits of agriculture and irrigation were the primary objectives for the construction of the Grand Coulee Dam, however neither the Colville or Spokane tribe realized these benefits as all of the irrigation water goes away from these reservations.
- Until recently tribes didn't receive any electricity or energy benefits, in fact, residents on the Colville Reservation pay some of the highest utility rates in Washington State which is provided by Bonneville Dam. The electricity to the local area is spotty. The majority of the benefits from the dam are seen in areas such as Seattle and Portland. Down-river cities and towns receive the benefits while the citizens up-river receive all the adverse effects.
- Construction of Grand Coulee has created several artificial lakes and reservoirs that attract millions of tourists and sports fishermen each year. The Colville tribe has increased costs because of the environmental damage and wild-land fire management impacts. Tourists and sports fishermen have access of hundreds of miles of pristine coastline along the Colville Reservation where they park their boats and camp at sites which are inaccessible except by boat and difficult to patrol—the non-Indian side is developed where they have to pay fees!

## 2.3 Contemporary Context of Native American Lands and Resources

As part of looking at Native American sustainable practices, it is important to briefly review the history of how the Euro-American settlers controlled land and resources on reservation lands. It is only within the last 30 to 40 years that Native Americans are beginning to make decisions on how and what resources to collect from their lands. The basis of Native American rights and their relationships to the U.S. government can be traced to treaties signed between 1778 and 1871 [53]. The BIA [53] summarized this as the

> "rights, benefits, and conditions for the treaty-making tribes who agreed to cede of millions of acres of their homelands to the United States and accept its protection."

Despite giving up millions of acres of homeland, the extent of the tribal customary lands was significantly reduced. This reduced the land area available where tribes could collect resources to generate income. The federal government took on the responsibility to make sure that Native Americans derived the income from their lands as part of their trustee responsibility. The Native American tribes, however, were poorly treated and their revenues stolen by the federal agency established to sell resources collected on reservation lands.

Tribes lost most of their lands and were put on reservations with other tribes that they historically fought wars. For example, the Confederated Tribes of the Colville includes the enrollment of the "descendants of 12 aboriginal tribes of Indians" (www.colvilletribes.com/2011_2012_ceds.php). There was not a rationale to why the U.S. federal government forced certain tribes to be located on the same reservation.

The other problem with being put on a reservation is that Native Americans lost a significant amount of the land area they had customarily used to collect resources. This land area was a fraction of the lands that each tribe had before the arrival of the European colonialists. When these treaties were made and tribes gave up rights to their lands, tribal members lost access to resources over large portions of the Northwest U.S. and Canada. This summary on the Colville Economic Development website (www.colvilletribes.com) describes this well:

> "Prior to the influx of Canadians and Europeans in the mid-1850s the ancestors of the 12 aboriginal tribes were nomadic, following the seasons of nature and their sources of food. Their aboriginal territories were grouped primarily around waterways such as the Columbia River, the San Poil River, the Okanogan River, the Snake River and the Wallowa River.
>
> Many tribal ancestors traveled throughout their aboriginal territories and other areas in the Northwest (including Canada), gathering with other native peoples for traditional activities such as food harvesting, feasting, trading, and celebrations that included sports and gambling. Their lives were tied to the cycles of nature both spiritually and traditionally."

Forests are an interesting lens to look at the challenges and changes in Native American ownership of resources located on the reservation itself. Rigdon [54] summarizes the importance of forests in delivering many of the cultural resources important to tribes:

> "Forests are a vital part of Indian communities due to the social, economic and cultural values the forests provide for tribal people. Reservation forest provides opportunities for economic development, employment and income, traditional hunting, fishing and food-gathering places and religious and cultural sanctuaries. Since time immemorial tribes have utilized and managed their forest for the resources they need."

Reservation lands are held in trust by the federal government for the benefit of tribes. Therefore, most of the history of Native American and U.S. federal government consists of the U.S. government regulating resource uses and effectively being the owners of resources located on reservation lands. "This trust responsibility is rooted in the Justice John Marshall 1830s court decisions involving the Cherokee Nation in Georgia. In Cherokee Nation v. Georgia, Marshall found that tribes existed as 'domestic dependent nations' within the United States." [54].

A number of policies were established by the U.S. government to regulate how Native Americans could use the timber found on reservation lands. These policies were driven by goals to make Native Americans into farmers (see section 2.1.1), so they would assimilate into the new American society.

It was not until the mid-1970s that Native Americans began to be able to manage their own resources themselves. At this time, they were able to develop their own policies on how to control lands and resources.

The following list summarizes some of the shifting policies that regulated ownership and Native Americans use of their forests [54]:

- 1873—U.S. v. Cook in stated Indians had no right to sell timber unless land was being cleared for agriculture and the trees belonged to the U.S. Rulings in Johnson v. McIntosh and Cherokee Nation v. Georgia reaffirmed that reservation lands and resources belonged to the U.S. government.
- 1887—General Allotment Act gave individual ownership of land to individual Indians. This resulted in the loss of reservation lands to non-Indians and the reason for the checkerboard patterns of land ownership found on many reservations.
- 1889—"Dead and Down Act"—gave tribes rights to collect dead trees and to sell them commercially. No live trees could be sold commercially.
- 1934—Indian Reorganization Act directed the Secretary of the Interior to make rules to manage reservation forest lands in a sustained yield man-

agement approach. This Act also verified that Native Americans were the owners of the lands and resources located on reservations.
- 1975—Self-Determination Act supported tribes "assuming the responsibility of managing many of the programmes once staffed by the BIA".
- 1976—Tribes establish the Intertribal Timber Council to coordinate the management of timber on reservation lands.

The extent of the land area held in trust by the U.S. federal government for tribes is large: 22.7 million ha [53]. The U.S. government administers about 326 Indian land areas as part of the reservation network. The BIA [53] described the powers held by tribal governments:

> "Tribes possess all powers of self-government except those relinquished under treaty with the United States, those that Congress has expressly extinguished, and those that federal courts have ruled are subject to existing federal law or are inconsistent with overriding national policies. Tribes, therefore, possess the right to form their own governments; to make and enforce laws, both civil and criminal; to tax; to establish and determine membership (i.e., tribal citizenship); to license and regulate activities within their jurisdiction; to zone; and to exclude persons from tribal lands."

Despite signing over 370 treaties with tribes [53], the U.S. federal government has not met many of its tribal fiduciary responsibilities. There is a history of tribes having to go to the U.S. courts to acquire revenues generated from tribally owned resources. This has not been an isolated case where revenues failed to be transferred to tribes.

The U.S. government obtained significant revenues from selling of tribal resources, e.g., oil, gas, grazing and timber, from reservation lands [55]. Recently, the U.S. government had to pay tribes billions of dollars in compensation for these lost revenues as part of the Cobell Case. The Spokesman article reported on April 12, 2012 that the federal government is paying 41 tribes to settle "a series of lawsuits brought by American Indian tribes over mismanagement of tribal money and trust lands" [55]. This is a long history of mismanagement, corruption and embezzlement of revenues generated from tribal resources when tribes were under the trustee responsibility of the U.S. government.

American tribes therefore have been unable to completely practice their model of sustainable practices where culture is the glue or core that determines the decision process, who collaborate on making decisions. Despite being robbed of their resources and lands, American tribal members maintain their practices and teach future tribal members how to maintain their culture (see Box 20).

Native American cultural practices can be compared to the ecosystem management and adaptive management concepts that are still being refined by western world scientists. For the western world these concepts are difficult to implement because the frameworks and tools to practice them are still evolving. The western scientific world is moving towards these paradigms but need to figure out how to link society in a credible manner to nature and its ecology. The nuts and bolts of a sustainability portfolio will be discussed next since it sets the stage for the holistic planning model used by indigenous communities even if they do not apply this term to their practices.

# A Lens on Cultures and Traditions of Indigenous Peoples and Local Communities

## Chapter 3

# Introduction to Folklore and Cultural Survival

*Our Book Leadership Essential:*
*The root of sustainable practices is maintaining ones link to nature through ones culture since culture creates diversity and diversity equals resilience.*

## 3.1 Western World Stories

The western European knowledge developed and is recorded in "turbulence" and not in stories passed down through the generations. Western Europeans can recount their family history going back to more than a thousand years. But these stories document more family lineages, e.g., who is related to whom. Native peoples know their family lineages but also have a rich history and many stories of relevant historical ecological information on nature. For the typical western world citizen, you do not look up your family lineage to learn about nature or gain insights into how to make competitive choices.

This is not a coherent body of traditional knowledge that facilitates decisions being made by the individual or by the community. The European descent people lack the long-term traditional ecological knowledge found with people who live for several thousand years on the same lands (see Fig. 3.1).

This does not mean that Euro-American descendants who live in the present day U.S. are not developing new traditions. They are evolving to become place-based to the lands but still retain the ancestry of their fore-parents who traveled from very distant lands to arrive in the Americas. But their knowledge is very recent and is not transmitted over seven generations like what occurs with Native Americans (also see the different views of nature in section 5).

There are few comparable ceremonies performed by most western European societies as what can be found with most native peoples. This does not mean that the western Europeans do not have symbols that are important to them. In

**Fig. 3.1** A Native American coyote telling stories to native children.
Figure credit: Ryan Rosendal.

Germany, Switzerland and Austria it is common to find a religious symbol placed in front of a field to protect the field and its grazing animals from evil spirits. Or, the changes in the seasons are still celebrated. Ceremonies are especially important in spring. Spring is a time to rejoice in the return of the cycle of growth in nature after a winter period of dormancy—a common ceremony in the Scandinavian countries. Most European ceremonies do not depict gods and goddesses or ancestral spirits who you need to communicate with to ensure that nature's abundance remains plentiful. Northern Europeans did have gods and goddesses a long time ago but lost these symbols after the spread of Christianity.

Because western Europeans do not recount stories that transmit nature knowledge or have a lesson to teach that your ancestors have passed down to their grandkids, their stories tend to be either "black" or "white" and its knowledge is developed in turbulence. These stories are very unlike a coyote (see section 1.4). Fairy tales told in the western world reflect this bias towards something being evil or good. Western European fairy tales tend to be dark and brooding stories where children are abandoned to fend for themselves, witches eat little girls or a wolf eats a cute little red caped girl. Or, the fairy tale tells a "good" story of a poor farmer's son accomplishing some great feat, even if it is only to kill flies or a big, nasty ogre. Mostly the poor farmer's son ends up marrying the King's daughter and lives happily ever after with her despite their differences in social status.

European type fairy tales are either black or white, e.g., right or wrong, and most are meant to scare the children into better behavior. There is not a very good message for children to hear since it suggests that the solutions are simple. There is only one solution possible. It does not suggest that there might be some elements of good and bad in every person or each situation. Or, every good story probably has some negative impacts on someone or some land recounted in the story. By contrast, the Native American equivalent stories are in shades of grey and meant to encourage children toward better behavior.

Similar to Native Americans, many other cultures also maintain inter-generational links to nature. Societies that have maintained this nature link are also those that have been able to maintain their culture in the face of intrusion from outside forces. As examples, the American Indians, Icelanders, Finns, and Indigenous Peoples of Indonesia maintained their cultures in the face of hundreds of years of extreme pressure to change their beliefs. These strong nature-based cultures have survived numerous battles against them. Native Americans had more than 500 years during which time they needed to survive after being conquered by the western Europeans. The Finns were dominated by the Swedes for over 550 years followed by over 100 years of dominance by the former Soviet Union.

Indigenous peoples in Indonesia survived under colonial governments for more than 350 years; Indonesia has been an independent country for the last 60 years. Until recently, indigenous peoples in Indonesia still had limited control and ownership of natural resources found on their customary lands. Regardless of the circumstances that played out in each location, these societies just mentioned have strong cultural roots that allow them to maintain their values and traditions despite the intrusion of other cultures on their societies.

Since the conquering countries make up the rules and determine or constrain the choices that indigenous people are able to make, this ability to survive and keep some semblance of one's culture is truly amazing. This is also the reason that today's global societies need to think about how culture helps societies to make ethical and environmental decisions, and to *survive*. These stories are worth briefly describing next.

## 3.2 Inter-generational Indigenous Cultural Stories

Many indigenous cultures survived and adapted to conquering onslaughts despite losing control and access to their lands and its customary resources. Sustaining their culture and ethics have been keys to this survival. Most indigenous communities, including American Indian tribes and Indigenous peoples (such as Dayak and Baduy in Indonesia), have legendary stories or folklore. They have a history of oration to recount these stories so they are not lost from the memory of future generations (see Fig. 3.1). Stories were used to entertain and pass on knowledge to families and a community—from children to adults—during many dark nights [56].

Culture and folklore are not just the past that can be relegated to dusty corners of libraries. Virtanen and Du Bois wrote [56]

> "easy to think of folklore as a thing of the past, . . . Even worse, it becomes difficult to image that people today can profit by studying such dusty antiquities of yore. Folklore becomes equated with the preindustrial community of old, the village or farm life so unfamiliar to many in the highly urbanized, industrialized Finland of today...Folklore is more than simply curious relics of the past, however. It is a dynamic and ever-present dimension of human experience. . . "

Virtanen and DuBois [56] (2001) wrote how folklore is a "broad-based inquiry into the intricacies of creativity and traditionality in everyday life." Finnish folklore is similar to the American Indian folklore in that it is enriched by its influences from the east and west. Indonesian folklore may be similar to other tropical countries', such as Brazil and Ghana, since these stories recount how the intrusion of different cultures is integrated into social norms. These stories also tell how societies adapt or do not adapt to technological change. They provide a lens on [56]

> "how people establish and pass on important values, norms and knowledge even in the face of constant change".

In his book *Coyote Finishes the People*, George [27] recounts a Colville story of the arrival of new people to the North American continent and how the coyote has to "finish" or prepare the native people to deal with new people (see Fig. 3.2). In this story, it is clear that the arrival of these new people will change their life forever. This story is similar to Indonesian folklore in that it recounts how societies adapt following the intrusion of a different culture and how they rewrite their stories. Folklore is alive and includes the new drivers of social change.

Since the new people were the western European colonialists, one would expect stark cultural contrasts between indigenous peoples and the Europeans. The European colonialists were the "intruders" into cultures that they were not a part off.

**Fig. 3.2** A symbolic Native American coyote.
Figure credit: Ryan Rosendal.

Folklore and stories are critical to any tool kit since they teach, entertain, and communicate with the listener regarding how to make ethical choices based on traditional cultures and knowledge. The American Indian coyote and Indonesian Dayak Tekena' stories are examples of the use of storytelling to provide a message.

The Native American coyote story is an example of the use of storytelling to provide a message (see Box 8). Many tribes of the Puget Sound have a salmon feast or first foods ceremony, where at the conclusion of the dinner they will return a carcass of salmon back to the ocean and they will also sing traditional songs and give oration during the ceremony.

During ceremonies, stories are told where the coyote or raven is depicted as one of the many main characters. Anthropomorphized animals are common in these stories and reflect the characteristics of those animals, e.g., cunning, strength, beauty, greed, etc.

Tribes have traditional knowledge and culture that are maintained by ceremonies so they are contemporary and not just interesting tidbits about an individual's family history and lineage. Ceremonies reinforce culture for tribes.

Native Americans, and other indigenous communities, have the cultural behavior and thinking that is maintained by their cultural resources and traditional foods. Plants are part of their ceremonies and traditional medicines (see Box 9).

**Box 8: A Coyote Story on Human Behavior towards Nature**

"...Coyote went as an invited guest to dinner at the home of Salmon. Salmon provided a fine feast, which turned out to consist of its own children. Before starting, Salmon said that they were not to cut across the bone, and at the end of the feast Salmon told the guests to return the salmon bones to the ocean.

The guests did so, and the salmon children reconstituted themselves.

Coyote, of course, thought this was a wonderful trick, so he decided to duplicate it.

He gave a big feast and served up his own children; at the end of the meal he asked that the bones be distributed into the ocean. This was done, but the coyote children were not reconstituted! Coyote himself was tricked.

Many of the modern Coquilles do not remember their ancestors doing what Salmon does in this story: returning the salmon bones to the ocean. But a few ... remember this ritual as part of some of the traditional salmon bakes on the beach.

And many remembered that their parents cut and served salmon in such a way as not to separate its bones. These practices show a concern with the regeneration of nature and the realization that human behavior and human attitudes affect nature's regenerative capacity.

Coyote, in this story, went through the ritual, but without the knowledge or the real capacity for regeneration, so Coyote had no success."

(from Hall [28])

**Box 9: Plants in Ceremonies and Medicines**

In addition to their uses in ceremonies, many plants have therapeutic value. For instance, when indigenous people of various tribes participate in ancient traditional sweat-lodge ceremonies, they often use different plants species such as sage, cedar or sweet-grasses. They apply these well-prepared dried plants to hot rocks within the sweat-lodge releasing a pleasing aroma which are known to have purifying and healing attributes that enhance the well being of the mind, body, and spirit. Furthermore, the steam generated from applying boiled water containing a tea onto fire heated rocks is known to promote purification, physical healing, and detoxify impurities of the inner body, thereby improving immunity; the tea is made of either sage, cedar, juniper, bitter root (cowes) or wild rose bush [57].

(Continued)

Indigenous people of North America utilized many plant species not only for their diet but for medicinal purposes. For example, *California sagebrush* (*Artemisia californica*) is commonly used by many tribes for antiseptic or cleansing reasons. Also, the leaves can be boiled in water and used as a tea for respiratory cold symptoms associated with coughing and it is known to ease menstrual cramps. Moreover, there are many remedies that indigenous people use in the Pacific Northwest; for instance, the Cowlitz Indians used red alder (*Alnus rubra*), containing salicyclic in the leaves and bark, for pain. Also, the Cowlitz used Douglas-fir (*Pseudotsuga menziesii*) and western red cedar (*Thuja plicata*) as cold remedies. The Cowlitz tribe use Oregon white oak (*Quercus garryan*) to treat tuberculosis [57]. The Klamath Rribe used mountain sage brush (*Artemisia tridentata*) as an antidiarrheal medicine [58]. The Washoe from Nevada used a sagewort (*Artemisia douglasiana*) and western juniperus (*Juniperus occidentalis*) as a pain killer [58]. Many of these vegetative species are gathered from tribe's local environments. Certain species are dried and ground into powder. The powder of plants leaves are made into teas then consumed; this is a practice that is still commonly used today when preparing these remedies.

(by Jonathan Tallman, Yakama Nation)

Both the American tribes and indigenous communities of Indonesia believe that if you do ceremonies right for nature, e.g., making sure that fish will return and rice will grow healthy with lots of grains, nature will continue to supply cultural food resources (see Box 10). Indigenous communities in Indonesia also have traditional knowledge and cultures that are maintained by ceremonies. These ceremonies keep their culture contemporary (also see Box 2 on ceremonies related to water).

**Box 10: Sundanese Kasepuhan on the Holiness of Rice**

Rice for Kasepuhan community of Indonesia is not only staple food. Rice is a holy plant; rice is a goddess, namely Dewi Sri. For Kasepuhan people, preparing rice field should be done carefully, with respect to Mother Nature. They only plant rice once a year and usually leave the rice plants in the field to obtain a second cycle of grain production. According to Uwa Ugis, an informal leader of the community, they will not harvest this second cycle of production:

"These grains are for wildlife, such as birds, insects, rodents, etc" (Personal communication to Asep Suntana in 2010).

(Continued)

By growing a second cycle of grains during each year for wildlife, claimed Uwa Ugis, the planting season next year will not be disturbed by pests and diseases.

The Green Revolution was introduced to Indonesia during 1970s. Its main purpose was to increase agricultural productivity. Although this was a massive program, it failed to force Kasepuhan people to use new varieties of rice. These new varieties of rice are much shorter (about 50 cm tall) than the local variety of rice, have awkward taste, and need plenty of chemical fertilizers and pesticides to grow. Since the new rice is shorter than the height of the local rice, the harvesting method is totally different than the traditional harvest method. In fact, it is not suitable for the ethnic beliefs, and socio-culture of the Kasepuhan people.

Since "without rice, Sundenese will die", Kasepuhan people conduct various ceremonies to make sure that the rice planting is not merely about growing rice, but also about communicating with the Goddess.

Ngabut [59] mentioned four functions for the oral stories spoken by the Kenyah Bakung tribe at Long Apan Baru, Bulungan district, East Kalimantan, Indonesia:

- Entertainment,
- Education,
- Instrument or media to communicate with magical sources, and
- Communication with non-tribal members.

This oral literature is divided into prose and poetry. The prose has five different functions, namely, *tekena'* (folklore), *ngidau* (mantra), *bon-bon usa talon* (to harvest honey at a very high tree), *tuba la'it tuba sanit* (to harvest fish), and *kelap ta' penyakit kini* (to ward of evil spirits).

Dayak Tekena' stories about natural resources provide assurance that natural resources surrounding each Dayak community will provide enough basic needs for all the community members. In several Tekena', Ngabut [59] mentioned about the close relationship between human (Dayak) and wildlife, such as hawks, wild chickens, and hornbills (Hornbills are the Dayak people's best friend).

For the American tribes and other indigenous communities like the Dayak, ceremonies make you stop and think about nature. It is not just a process to use scientific knowledge to make decisions but thinking about the other cultural values that nature provides (see Box 10). Many Native Americans/tribes will talk about respecting the gifts of the Creator so that you will always have this resource. Tribes have a teaching that the deer or roots or berries will show themselves to you if you give them respect. This is a teaching in the Colville longhouse. Rodney Cawston heard the Nooksack Tribal Chair say almost the exact teaching during one of their meetings at their tribal council.

In the section 3.2.1, Rodney describes how he grew up learning the cultural practices of his tribe.

### 3.2.1 Learning Nez Perce Culture while Growing Up as Remembered by Rodney[1]

In 1855 Chief Joseph's father, Old Joseph, signed a treaty with the U.S. that allowed his people to retain much of their traditional land in the Wallowa Valley in Oregon. In 1863 another treaty was created that severely reduced the amount of land, but Old Joseph maintained that the second treaty was never agreed to by his people. The second treaty, signed by Chief Lawyer in 1863, significantly reduced the area of the tribe's reservation by 90 percent, ceding away the traditional homelands of many Nez Perce bands. Although, this was done without their consent, Chief Joseph and his people prepared to move onto the Reservation at Lapwai when a skirmish broke out and this started the Nez Perce war. For over three months Chief Joseph and his people fled 1,883 kilometers over Oregon, Washington, Idaho, Montana, and Wyoming hoping to make it to Canada. On October 5, 1877, approximately 48 kilometers miles from the Canadian border and after months of fighting, Chief Joseph surrendered. Eventually, he and many of his people were sent to "Indian Territory" in what is now Oklahoma, where many died from malaria and starvation. Chief Joseph unsuccessfully appealed to the federal authorities to be able to return to their traditional homelands at Wallowa. In 1885, The Chief Joseph Band was sent to the Colville Reservation in Washington.

Descending from the Chief Joseph Band of Nez Perce, I can recall hearing many orations from my Elders at the Nez Perce Longhouse on the Colville Reservation. On the Colville Reservation at Nespelem there is an old log hewn home that was built by hand by the Nez Perce and Palouse people who first moved onto the reservation that served as our longhouse (see Fig. 3.3). The photo at the right is of Frank Andrews in front of Old Nez Perce Longhouse taken in 1994. At this place, Nez Perce people would regularly gather for traditional ceremonies, Sunday services, first food feasts, funerals, memorials, weddings, birthdays and other celebrations. Nez Perce people gathered here until this building was replaced in the early 1970's by the present day longhouse.

When I was a child, I remember being at one of our root feasts at the old longhouse, the tables were cleared and I listened to many of the Elders speak about why it was important for us to learn our language, our songs, our stories,

1 Confederated Tribes of the Colville.

**Fig. 3.3** The old Nez Perce longhouse, 1994.
Courtesy: Washington State Folk Arts.

and our culture and to pass this down to our children. They said that our ancestors made a very difficult decision to move here into their enemy territory. But they did this to keep our religion and culture.

> "We believed in our own Hunyewat [God, or Deity]. We had our own Ahkunkenekoo [Land Above]. Hunyewat gives us food, clothing, everything. Because we respected our religion, we were not allowed to go on the Nez Perce Reservation. When we reached Wallula, the interpreter asked us, "Where you want to go? Lapwai and be Christian, or Colville and just be yourself?" No other question was asked us. That same had been said to us in our bondage after knowing we were to be returned from there. We answered to go to Colville Reservation.
>
> On the Colville we found wild game aplenty. Fish, berries, and all kinds of roots. Everything so fine many wanted to remain there, after learning that Wallowa was not to be returned to us. Chief Moses advised Joseph to stay. The Indians were good to us. Gave us horses, and good salmon at Keller. It was better than Idaho, where all Christian Nez Perces and whites were against us." [60].

During the month of July, each year since the Nez Perce were moved onto the Colville Reservation, the Nez Perce people would set up their tipis and camp together for our annual celebration (Fig. 3.4). This was always a ten-day celebration which began with a memorial horse parade, memorials, and a dinner.

The memorial parade originally started as a commemoration of all the Nez Perce people who were lost in the war. Families would also use this time as a memorial for lost family members by leading a rider-less horse through the parade. Many of the families would have their own private memorials and other ceremonies in their camps throughout this time. Later on, many of the other tribes on the Colville Reservation joined in on the festivities of song and dance and traditional games. These celebrations still take place and although much of the dancing has evolved as having many of our traditional pow-wows, the intent of the celebration is still the same.

**Fig. 3.4** Nez Perce Celebration July 4th, Nespelem, Washington, CA. 1900–1910. Courtesy: University of Washington Libraries. Special Collections Division.

Our Elders often tell us that our religion and culture gives us our individual and unique identity. The traditional religion of the Chief Joseph Band of Nez Perce on the Colville Reservation has never ended. Throughout our history at Wallowa, at Oklahoma and now on the Colville Reservation, our Elders handed down our religion and culture through many events and ceremonies each year. Our culture shapes our community through the language that we speak, the Nez Perce names we carry, the traditional art forms we have, the way we gather and respect our traditional foods, and by learning and teaching our stories, songs and dance. In many ways, we are taught that the importance of culture cannot be stressed enough as it gives us our world views and is an integral part of our living.

Despite Native Americans and indigenous communities continuing to practice their cultures and traditions, their practices are frequently not accepted by the western world. The western world does not understand, and does not want to understand, these long cultural and historical traditions. They do not accept them as being civilized since they have established their own regulations on how resources should be collected and consumed. A good way of controlling native peoples lands is to set up your own rules on what they can practice.

It also is a problem that the lands and resources owned by native peoples continue to be coveted by industrialized countries. Native peoples continue to lose their customary lands, e.g., land grabbing, as part of government strategies to develop regional economies. Some indigenous peoples have been killed when they tried to prevent others from stealing their lands. Henry Saragih, Chairperson of the Indonesian Peasant Union (SPI), reported 20 deaths of farmers during agrarian conflicts in 2011. There were over 120 cases of agrarian conflicts alone in Indonesia in 2011.

The next story describes how an Alaskan native faced prison time because of killing a moose when it was not an acceptable hunting season (see section 3.3.1).

## 3.3 What Does It Mean to Be A Traditional Ecological Practitioner?

> *"...I know it's a widespread assumption in the West that, as countries modernize, they also westernize. This is an illusion. It's assumption that modernity is a product simply of competition, markets and technology. It is not; it is also shaped equally by history and culture."*
>
> ~ Martin Jacques [61] ~

### 3.3.1 Break the Law When Practice Culture

The western world legal translation of what is an acceptable cultural practice continues today. Native Americans have been convicted in courts of law for practicing their cultural traditions since it was contrary to western laws that set aside a particular time for an activity to occur.

The summary in Box 11 shows the conflict that can occur when the western world has established specific times for hunting to occur. This was not compatible with the traditional practices of native peoples. In this case, killing a moose for a funeral potlatch broke the law since the hunting season was restricted to particular times of the year (//law2.umkc.edu/faculty/projects/ftrials/conlaw/frank.html; accessed Oct 8, 2012). This story from Alaska reflects how a Native Alaskan faced prison time for following a cultural practice that had occurred for hundreds of years. This story is described in Box 11 (Carlos Frank, Appellant, v. State of Alaska, Appellee; Supreme Court of Alaska; 604 P.2d 1068; December 21, 1979):

**Box 11: Native Alaskan in Prison for Practicing Funeral Ritual**

"In October of 1975, Delnor Charlie, a young man from Minto, died. Immediately preparations were made for a ritual that had been performed countless times in Minto and other Central Alaska Athabascan villages. It is called the funeral potlatch, a ceremony of several days' duration culminating in a feast, eaten after burial of the deceased, which is shared by members of the village and others who come from sometimes distant locations... Delnor Charlie's burial, as is traditional, was delayed until friends and relatives living elsewhere could reach Minto and until the foods necessary for the potlatch could be prepared.

With the food preparation under way, Carlos Frank and twenty-five to thirty other men from the village formed several hunting parties for the purpose of taking a moose. It was their belief that there was insufficient moose meat available for a proper potlatch.

One cow moose was shot, which Frank assisted in transporting to Minto. Some 200 to 250 people attended the final feast.

A passerby took note of one of the hunting parties and reported it to state officials, who investigated and subsequently charged Frank with unlawful transportation of game illegally taken. The season for moose hunting was closed and in any event there was no open season for cow moose in 1975."... Frank was thereupon convicted and sentenced to a forty-five day jail term with thirty days suspended, a $500 fine with $250 suspended, one year probation, and a suspension of his hunting license for one year."

Even under less drastic circumstances, the lack of knowledge about native people's traditions and culture is quite apparent. This happens even when western people want to show respect for native traditions.

In Washington State, non-Indians may still have difficulty understanding what it means to practice cultural practices. A typical western world approach is to acknowledge the importance of Native American practices but not to understand what these cultures are (see Box 12).

Native Americans, and other indigenous communities, have the behavior and thinking that bound their knowledge base and make them ideal *Global Sustainability Managers*. The cultural traditions and management practices used to maintain cultural resources and traditional foods are described in Box 13.

Once we understand how culture is maintained through several generations in society, it is important to think about how a Native American maintains his or her Indianness. It sounds simple. You need to continue practicing your traditions and ceremonies. If you want to become a traditional ecological practitioner, you need to start following their practices. You could even start to read the right fairy tales or listen to the appropriate cultural stories. Unfortunately, it is not this simple and we know that.

**Box 12: Washington's Centennial Celebration 1989 by Rodney** (Confederated Tribes of the Colville)

In 1989, Washington State celebrated 100 years of statehood. This was also the year when twenty-six sovereign tribes in the State of Washington signed the Centennial Accord, which is an agreement that provides a framework for the governor and Washington's tribes to promote tribal self-sufficiency, work together to achieve mutual goals and to develop stronger tribal-state relationships.

During this year, the Colville Confederated Tribes opened up a museum and gift shop and with the grand opening, the tribe held a reception for the governor, state and federal legislators and other dignitaries. This event was in recognition of the Washington State Centennial Committee's support of the Museum and recognition of the Centennial Accord. Many preparations were made, a local ball room was reserved, caterers were hired and flowers were ordered. Some of the tribal council felt that it would be important to serve traditional foods at the reception. The event was elegant with crisp white table linens, polished silver, crystal glasses and a traditional artist played flute music to create a pleasing and relaxing atmosphere. Young men and women were traditionally dressed and they greeted the guests by pinning a culturally significant corsage on everyone as they entered.

A very respected tribal elder agreed to provide traditional foods for the meal. This elder brought in the foods and arranged them along with the rest of the foods on the buffet tables. These traditional foods were presented in re-used cool whip and other plastic containers and enamel pots. The dishes were covered with tin foil. Many of the guests were very squeamish about taking or eating any of these foods. This elder took the governor by the arm and led him to the traditional foods and said: "You are going to try everything". She filled his plate with wild roots, berries, smoked salmon and dried deer meat. He ate everything and gave her compliments for the meal. To this elder, providing such a fine meal was an action of great respect. Native American people have reused and recycled resources way before these efforts were the popular prevailing practice. The event was regarded successful despite cultural differences or the misunderstandings and lack of common respect.

**Box 13: Examples of Traditional Ecological Knowledge and Culture**

For thousands of years, indigenous people of North America utilized several landscapes and developed a fundamental way of life as agriculturists, hunters, and gatherers by burning many landscapes. Many of the tribes in the eastern U.S. planted an abundance of crops such as corn, squash, gourds, beans and many other plant species [62]. Therefore, indigenous people made use of the many different ecosystems.

Native Americans established many harvest and subsistence systems to meet their requirements of nourishment and support their welfare. One important way that indigenous people enhanced their well-being was through their diet and nutrition. Native Americans are considered to be the first ethnobotanists of the Americas.

The historical use of fire by Native Americans is apparent within the prairie grasslands, montane meadows, oak woodlands, and conifer forests [63]. Their approach to using fire was to achieve specific objectives, and to improve desired vegetative communities. For example, Native Americans used fire to enhance edible food sources such as camas.

The indigenous tribal people applied fire onto prairies and meadows to encourage future plant productivity for wildlife, especially for foraging and browsing animals such as deer and buffalo. Furthermore, archaeologists suggest indigenous peoples adapted to ecosystems of North America mainly by practicing hunting and gathering. Hunters utilized sharp arrow heads for spears, bows, and arrows as tools for hunting. Gatherers developed methods to gather and preserve foods from North America for close to ten thousand years.

Over the course of nearly ten thousand years, indigenous people of North America established a relationship with the natural resources: water, vegetation, prairies and forests. Furthermore, the Indians made extensive use of fire to shape the vegetation and the ecology of forests and prairies.

They employed fire to enhance the growth of foods such as camas or to attract deer and elk to the rejuvenated vegetation following a burn. They figured how to utilize fire through teachings of knowledge past down from generation to generation to shape and enhance vegetative processes within many ecosystems. This sometimes is referred to as traditional ecological knowledge (TEK) or "the knowledge base acquired by indigenous and local peoples over hundreds of years through direct experience and contact with the environment," and developed for thousands of years [63].

(by Jonathan Tallman, Yakama Nation)

Wendell George has been asked by other native peoples on how to maintain Native American cultures and traditions. Even native peoples find it hard to continue to practice their traditions when facing all the marvels that modern technology has to offer. It is easy to be seduced to forget long traditions when technologies provide instant gratification. Finding a balance between the global and local environment and culture is critical. You cannot ignore the global and only focus on the local cultures. In fact, a practitioner has to become like the coyote as shown in the illustration—the coyote has a worldly perspective but at the same time can recite local folklore and use it to think through decisions that need to be made (Fig. 3.5).

**Fig. 3.5** An image of the coyote looking over the world and using his local nature culture to make natural resource decisions.
Illustration courtesy: Ryan Rosendal.

Wendell George provides his perspective on maintaining one's Indianness in the next story.

### 3.3.2 Indian Spirituality[1]

We have come full circle. American Indians have long maintained that we are one with nature. In recent years the rest of the world has increasingly come to

1 By Wendell George, Confederated Tribes of the Colville. Books written by Wendell George: *Coyote Finishes the People* (2011) and *Last Chief Standing: A Tale of Two Cultures* (2012).

the same conclusion. Many books have been written about the Gaia hypothesis. Gaia was the Greek Goddess of Earth. Indians called her Mother Earth and the Sun is called Father. The Gaia hypothesis describes the earth as a living system.

The dominant culture is too preoccupied with the physical side of life. As a group they tend to think they are something separate from creation because their culture lost awareness of the earth as a living system. This thought process is the root cause of many of the problems today. Indian spirituality is based on the premise that we are one with nature or one with Haweyenchuten (Wenatchi for Creator-God).

Yet it is difficult for most Indians to maintain their culture. We have to rekindle our Indian belief that we are one with all nature, including the stars and galaxies, not just trees and animals.

Even though some Indian families are doing a good job with their children, they are finding it more and more difficult to instill "Indianness" in them because they are confronted daily with non-Indian activities, beliefs, and attitudes.

This can be dramatically demonstrated with the Bev Doolittle drawing of what at first glance looks like a forest scene. On closer inspection several faces of Indians show up. They are hidden among the trees just like modern day Indians are hidden among the multitudes. All of us have to look at life just right to see the Indian in us.

When Europeans first came over to America they began teaching Indians to live like Europeans. Indians were encouraged to adopt the practice of owning private property, building homes, farming, learning European languages, religion and customs.

This massive acculturation effort was performed without understanding Indian culture. Europeans didn't realize that much of the native culture was the same as theirs but derived by looking at the world from a different perspective. And, more importantly, that viewpoint revealed a better understanding of the world.

It is time that the dominant society becomes acculturated to the native culture. This must start with an understanding of what Indians really are. As this understanding grows maybe some of the harm to the world can be undone.

History has shown that many things have been taken away from Indians so we are becoming the invisible American even though we are the original Americans. One example is when Stanford University decided to change their mascot from "Indians" to "Cardinals". This started a national movement where many have eliminated any reference to Indians. It was thought to be a do-good change which was caused by the collective conscience developed from observing the genocide of Indians.

As early as 1925, Zane Grey wrote a book called the "Vanishing American", which was about Indians disappearing. Indians are not disappearing but they are becoming more invisible. This national conscience is misplaced because Indians still want to be recognized. In 1930 the name Indians was brought to

Stanford by Pop Warner who previously coached at the Carlisle Indian School in Pennsylvania made famous by the World Olympic Champion, Jim Thorpe. It was an honor to have Stanford called Indians because it recognized Indians for who they were.

Harmony can be achieved by integrating oneself with nature, trees and animals in the woods. Many people today have pets and become very close to them and achieve harmony. This harmony can also be applied to families, countries, and the earth as a whole.

The European western culture has interrupted this harmony with their concepts of land ownership and their driving need to conquer and control the earth (see section 5.1). Indian spirituality is present when you "know" things are right and in harmony. Today it would be called intuition, but to be spiritual it must be based on spiritual not physical values.

Some people from other cultures have recognized the tie to nature and have studied it in depth. Unfortunately, they have not spent much time with Indians and have glossed over or ignored the spiritual side. Their investigations are mostly a condemnation of the physical harm that civilization has caused to the earth.

We have to look behind those physical results to understand the root cause. Let's compare Indian values with others. Indians tend to be right brained:

| Use Left Side of the Brain | Use Right Side of the Brain |
|---|---|
| 1. Order, logic, math, details | 1. Perception, creativity, art, generalize, intuition |
| 2. Science, reading, facts, writing, planning | 2. Spiritual, music, dance, imaginative, humor |
| 3. Systematic | 3. Symbols |
| 4. Linear | 4. Non-linear |
| 5. Goals | 5. Feeling |

This results in cultural differences:

| Western Culture | Indian Culture |
|---|---|
| 1. Competitive | 1. Cooperative |
| 2. Aggressive | 2. Emotional |
| 3. Personal goals important | 3. Group needs more important |
| 4. Future time orientation | 4. Orientation to present |
| 5. Control of others | 5. Control of self not others |
| 6. Trial and error learning | 6. Participation after observation |

Which results in different value systems:

| **Western Culture** | **Indian Culture** |
|---|---|
| 1. Only humans have a spirit | 1. Everything in nature has a unique spirit |
| 2. Nature is to be conquered and tamed | 2. Humans should live in harmony with nature |
| 3. Scientific methods used for healing | 3. Healing by being in harmony with nature |
| 4. The more you own the better you are | 4. Sharing and generosity are greatly valued |

These traits describe original tribal people and can be confused today because most Indians have been integrated into the dominant society. However, the tendency is still there and will show up from time to time.

Some consider Indians to be the keepers of the earth or as aboriginal time's keepers of the fire. But how can such a small population influence the vast majority? It can but only if supported by enough influential people. We must reach a critical mass to cause the butterfly effect as described in the introduction (see section 1).

Ervin Laszlo wrote in his 2006 book, "A critical mass of people in society must take an active role in creating a new civilization. They must shed obsolete beliefs, adopt a new morality, envision the world as they would like to see it and evolve their consciousness. This can be done by moving from the obsolete Industrial Age of conquest, colonization and consumption to connection, communication, and consciousness." Ervin Laszlo is founder and president of the Club of Budapest, which is an informal association of people in fields of art, literature, the spiritual domains, the business world and civil society at large.

Psychologist Carl Jung and physicist Wolfgang Pauli concluded that our mind seems to be subtly but effectively connected with other minds and other bodies. This connection lies beyond both psyche and physics. Many scientists have concluded that the universe is a living organism and that we are all connected at the quantum level.

Sir Julian Huxley and Pere Teilhard de Chardin described this spirituality evolution in people in the 1959 book, *The Phenomenon of Man.* They thought of the universe as being one gigantic evolutionary process. It is a process of becoming and attaining new levels. In addition, it is being led to cultural convergence. The Eastern culture and Western culture are merging together to form a unification of world thought. It is tied to the psychic energy that only the human being is capable of processing. Mankind facilitates exchange with the universe. Omega is the final state of the process of human convergence. The human is psychically tied to the earth, which becomes a true microcosm of the cosmos. The human has crossed the threshold of self-consciousness to a new mode of thought. The human personality trends towards more extreme individualism and more extensive participation in its own development. This new level will lead to new patterns of cooperation among individuals, cooperation for practical control, for enjoyment, for education and new knowledge. It will result in love, goodwill and full cooperation among people the world over.

Both Laszlo and the Mayans agree that the transition will take place slowly. The Mayans from the Yucatan in Mexico say the world change started in 2007 and will continue at least until 2015–2018. Laszlo describes a new world view in 2025 of a holistic civilization.

To develop a holistic civilization that is based on connection, communication, and consciousness we can use the American Indian model. It is a simpler lifestyle but more inclusive.

Even though most tribes didn't have a written language they generally agreed with the Iroquois League of Nations constitution recorded in 1570 (see section 10.2). Their constitution included initiative, referendum, recall, women's suffrage and a Bill of Rights. Ben Franklin was so impressed with it that he recommended it in his Albany Plan of 1754. The U.S. finally adopted much of it in a watered down version. Later they adopted a Bill of Rights that included more of the Iroquois constitution. Many issues became evident during these discussions.

For example, women were not allowed to vote until 1920 when the nineteenth amendment was approved by the states. Indians always included women in their organization. They were actually more important than the men for the survival of the tribe. The Iroquois women trained the boys to be leaders and to be responsible for their families and tribe. This was an everyday experience and became built-in to the psyche as each child was growing up.

Initially in the United States only land owners could vote which of course gave them the advantage to keep all legislative actions in their favor. George Washington was a big landholder and the richest President of the U.S. This mind set has continued into the financial industry.

In all the time that people were foragers, they never developed more complex, larger scale societies. There is a connection between agriculture and the complex society that has evolved (see section 2.1.1).

In a About 11,000 BC agriculture started in Southwest Asia, specifically Mesopotamia, and spread throughout the world over the next few thousand years. Since agriculture provides more food per unit area of land compared to foraging it allowed more people to live in a given area. This satisfied the social desire of humans. However, it takes more labor to farm an acre of land than to collect the world foods that are naturally there. Although it may take twice as much labor to farm an acre you do not get twice as much food from that acre. But then as the population grew there soon became too many people for the wild resources to support in any given area. So farming became the only alternative. It soon became apparent that farmers have less free time than foragers. But that was overlooked.

Agriculture made it possible to overproduce and store the surplus. This created a new problem for the evolving society to face. For example, should each farming family store and control its own surplus, or does some or all of it get stored in shared, communal arrangement. Before they could live in larger groups they had to solve the problems it would create.

These complex social organizations required the development of institutions to settle conflicts. The permanent locations of stored food surpluses were easy targets. They were attacked first by individual thugs, then aggressive armies and finally by the Internal Revenue Service. So power hierarchies came into being.

As populations grew so did the desire for more agriculture. More agriculture required more land so land became a commodity. The initial handful of raiders gradually became large armies who conquered large communities for their land. These armies were ruled by kings who came to power by divine fiat.

A maxim that is a good business practice today is to delegate decisions and responsibility to the lowest level at which they are effective. This is in exact opposition to the top down approach in today's society which is derived from the military mind set.

This is demonstrated in the activities of the American Indian tribal chief. He was not the supreme commander as in the dominant society. He was nominated or appointed by his tribe because he could make good decisions but he did not make them in a vacuum. He discussed the issue as long as necessary to get the consensus of his people. His decisions were usually unanimously approved.

On another issue the Chief didn't take part of everyone's belongings so he could give it to someone who he thought needed it. On the contrary, he gave away his own to help his people. He was sometimes the poorest one in his tribe because he gave everything away. This attitude trickled down to everyone in the tribe. Giveaways were a custom especially when someone passed away. They each helped each other directly and voluntarily. There was no law that said they had to do it. It was true government by the people. The governing body had no power unless specifically given to them by the tribal members. It was the opposite of a dictatorship. If a chief became selfish he was soon replaced (recall). The need for a governing body was only for major problems such as warfare or famine. Tribal members were left to their own devices but tempered by their upbringing which required them to do what was best for each member of the tribe. Independence was prime. Any law created would be a reduction of their independence.

The paradigm shift from the physical domain to the spiritual domain will not happen unless it is broadly accepted by the community. To do this we need to jolt all of humanity into a spiritual awakening. The key to this is to increase the use of the right brain. Indians are more adaptable to this because it is more natural to them. Europeans have to be taught to do this. Our existing society has caused 90% of our people to use the left brain in its thinking. It is inherent in our education, business, health care, and government systems.

We live in a left-brain world. We are interested in the exterior, material environment and involved in activities, motion, logical thinking and the understanding of time and space. The right brain is more interested in internal than external. It is passive and meditative. It guides its thinking by feelings and hunches. It is creative, intuitive, and instinctive. This seems out of step with the material world. That is why conventional methods of education have virtually ignored it. The purpose of those methods is to prepare us to survive in the material world.

So the right brain does not figure prominently in our modes of conducting government, international relations, politics, business or health care. Many have recognized that and have created a number of self-help programs. A very effective program that has trained over 10 million people since 1966 is the Silva Mind Control Training developed by Jose Silva, a Native American of Mexican descent. Their techniques are directed at helping to emerge a world based on cooperation rather than competition, an affirmation of the human spirit and to demonstrate that all humanity is connected. This is the gateway to spiritual harmony.

The approach is simply to enter the alpha level (brain waves operating at 4 to 7 cycles per second) by relaxing your mind. They use several techniques to enable the right brain to solve problems. These techniques are similar to the American Indian sweat houses, winter dances or the search for one's Sumax (spirit power). This approach has helped many people and organizations in science, education, politics and health care to improve their approach to life. People have become more humane and exhibit more respect for all forms of life. Natural resources are used more efficiently. Labor and management communicate and work together for the common good. People in the finance and marketing arenas and the government leaders become more honest and straightforward. If this continues we will eventually reach the critical mass needed to create a new civilization based on spiritual needs.

Maintaining one's culture is also dependent upon a person keeping one's language alive or relearning it if it has been lost from one's vocabulary. Language is important in ceremonies and is used to practice culture. You cannot do ceremonies correctly if it is spoken using the language of societies that intruded on your culture. Native Americans had hundreds of different languages that they spoke before 1492. Most indigenous communities globally are working at relearning languages that they lost when the European conquests began. A brief discussion is provided next to make the reader think about how languages are important for maintaining one's Indianness.

### 3.3.3 Native American Languages[1]

Scientists generally agree that humans have developed approximately six to seven thousand spoken languages. It is estimated that thousands of these languages were spoken by indigenous people in North and South America prior to the first contact with Europeans at the beginning of the 11th century. The indigenous languages of the Americas had widely varying demographics, from some which had millions of active speakers, to those that had only several hundred speakers. Some, but not all, indigenous cultures of the Americas also developed their own written languages, essentially visual symbols recorded on some type of medium to represent ideas that can be expressed in a physical or verbal form of communication.

Historical explorers who came to North America had widely varying attitudes towards Native American languages. In some locations, missionaries learned local languages in order to preach Christianity and other religious doctrine to the indigenous people in their own language. Religious zealots often sought to convert Native Americans to Christianity as part of their efforts to conquer them and bring them under control. However, it was much more common for the Europeans to vigorously suppress Native American languages. As a result, knowledge and use of Native American languages greatly declined. By the 18th and 19th centuries, the Spanish, English, Portuguese, French, and Dutch, brought to the Americas by Europeans and others, had become the official or national languages.

It is well-documented that European and other explorers destroyed books and other documents maintained by indigenous people they encountered. The majority of the written Native American records were destroyed as part of campaigns to conquer or control them. The historical written records that exist are currently under threat of being lost or destroyed. Funding for archives is severely limited. Petroglyphs and pictographs, which are sometimes the only surviving examples of specific Native American languages, are being lost to environmental degradation and tourist's graffiti.

Although many indigenous languages have become critically endangered and many have been completely lost, some are still rigorously spoken. In the United States, Navajo is the most spoken Native American language north of the U.S. – Mexico border, with over 200,000 speakers primarily in the Southwestern United States. The number of Navajo speakers continues to increase. A number of bilingual immersion schools are established in the Navajo Nation to preserve and promote usage of the Navajo language. Navajo was used by the Navajo Code Talkers during World War II to transmit secret U.S. military messages, which neither the Germans nor Japanese ever deciphered.

1 By Melody Starya Mobley, Cherokee.

Today, Native Americans, the federal and state governments, universities, nonprofit organizations, the National Science Foundation and others are working hard to preserve and revitalize the use of indigenous American languages. Some scientists and others view Native American languages as endangered American resources, another treasured resource that is critically in need of protection. In this sense, protection means and includes research, preservation, restoration, and passing these languages along to future generations, especially children, and others beyond tribes of the languages' origins. These languages often express ideas about Earth, its natural resources and the management and value of these resources very different from European and other thinkers.

Historical Native American languages can introduce us to the "foot soldiers of conservation." These days, Native American languages are most often only heard at ceremonial events with little if any participation by other than Native American people of the language's origin. The Ojibwe language was once spoken in the region surrounding the Great Lakes in North America. Like the majority of Native American languages it is endangered of being lost.

Professor Mary Hermes of the University of Minnesota, Duluth Education Department, is working hard to keep the Ojibwe language alive by bringing together tribal elders and video technology in a "Transparent Language System." With important support from the National Science Foundation, Professor Hermes videotapes fluent Ojibwe language speakers in everyday situations then elders and others help transcribe these short films.

The result is a treasure trove of language lessons from master Ojibwe language speakers that are archived for preservation into the future and current use by students and linguists everywhere. Professor Hermes and the team brainstorm themes, situations and stories to preserve then videotape master speakers, including elders, improvising the language within them. The goal is to record the Ojibwe language in ways that may not be grammatically correct but it is the way the language is actually spoken. Ojibwe elder Ruby Boshey is excited about the project because she notes that as elders die the Ojibwe language is being lost.

This is a common issue in the majority of Native American communities. Professor Hermes began by building relationships and trusts within the Ojibwe community, and speaks Ojibwe herself. Dr. Hermes also organizes and holds workshops and immersion classes, and develops a series of interactive educational games. Hermes stresses the benefits that saving Native American languages has major payoffs beyond a specific tribe. "There's 10,000 years of human evolution and knowledge in that language," she emphasizes. New words are routinely added to the Ojibwe vocabulary, as occurs with all "living languages".

The Ojibwe language has two-thirds verbs and is considered one of the most difficult spoken languages. It is easy to visualize how this language captures or contains descriptive, complex thoughts, unique ideas and other information, how much it could teach all of us, and what is being lost every day or has already been lost. This is critical information we cannot afford to lose especially with the threats globally on natural resources and climate change. And, Ojibwe is the only one of a plethora of Native American languages.

It may not be instantly apparent how much has been lost or what are the ramifications of those losses, especially when early explorers and others destroyed all aspects of the lives of Native Americans in insidious ways. To further reduce the ability to communicate with Native American languages, it was common to force the removal of Native American children from their biological parents and extended families and place them with non-Native American foster families or worse, place them in group locations. Children were not allowed to wear traditional Native American clothing or dress in ways normal for them when they were forced to attend English schools. Native Americans were forced to eat foods foreign to them. Digestive problems, anemia, and undernourishment were prevalent, in addition to the loss of cultural associations with foods, for example, personal power and strength and ability. Knowledge of the medical use of native plants and foods was lost. This further weakened, lowered the health and destroyed the immune systems of Native Americans, and the majority of this knowledge has been lost to us today, perhaps impacting modern health and science.

There are additional current positives in the acceptance or valuing and efforts to preserve Native American languages and traditional practices and rituals besides the example outlined in Professor Hermes work. The irony cannot be overstated because a real positive resulted from the involvement of the current Catholic Pope Benedict the 16th when he canonized seven Catholic saints in Rome, Italy, on October 21, 2012. One of the canonized saints, for the first time in history, was a Native American, Sister Kateri Tekakwitha, a Mohawk woman from New York. She is the first Native American to be designated a "Saint". According to the Vatican, prayers to this Mohawk woman, Sister Kateri, were responsible for curing a 6-year old boy, "Jake", from Washington State who suffered with infections of flesh-eating bacteria that severely disfigured him and threatened his life. The little boy's family pastor suggested praying in the name of "Blessed Kateri Tekakwitha" because when she died in the 1600's it was noted that her scarred and disfigured face was healed and made beautiful by God.

It is believed that the intervention of Blessed Kateri led to the healing of Jake, who is healthy and 12 years old now. That miracle is why "Blessed Kateri" was designated the first Native American saint. Alma Ransom, former Chief, St. Regis Mohawk, and a Catholic herself, participated in the canonization ceremony for Kateri. Former Chief Ransom stated that although they had waited over 300 years for Kateri's gifts to be recognized officially outside the Mohawks, people from across the United States participated in the ceremony for Saint Kateri, some in traditional Native American dress.

Journalist Martin Savidge, CNN-America, asked former Chief Ransom if in any way in her mind there was a conflict between Native American spiritualism and the Catholic faith. Ransom replied that the tradition in the Mohawk culture is "our Creator is your God". Ransom went on to state that in her time Kateri "had a tradition where she knew her Creator very well and all the rules that go with it—loving and respecting the elders, loving Earth as an environment. She had all these things, an appreciation of all of God's creation," Ransom said, "When she became Catholic the Creator, God, became Jesus... And she laid a wonderful example for us to be able to live in both cultures." Describing the ceremony in Rome, Ransom said, "We sang and prayed in Mohawk. We still speak the same language, say the same prayers that Kateri Tekakwitha prayed when she was a young woman."

The religious culture that attempted to thoroughly destroy Native Americans and demonized them now makes them saints.

Practicing traditional cultures or ecological knowledge does not take just one form. Indigenous communities have variable practices that keep them embedded in their cultures. It is important to realize that there is not one model that we should adopt to humanize sustainable choices. Whether an approach should vary depends on the location of a community and how they have historically responded to their changing and dynamic environments. A framework provided in section 4 provides our view of the opportunities and constraints to practice culture. Indigenous communities in Indonesia will be used to highlight how different indigenous communities balance their local cultures with the constraints and opportunities of surviving on their lands.

**Coyote Essentials**

- Diverse cultural practices are adapted to specific locations.
- Cultures and traditions are not irrelevant even for technologically advanced societies.
- Language, ceremonies and taboos keep cultures alive and linked to nature.
- For humans to take better care of the earth, art and technology need to have equal emphasis.
- Technology by itself will not produce sustainable nature practices but a balance between technology and spirituality will.
- Traditional knowledge does not provide solutions for a specific problem but an approach for humans to follow that includes nature and humanizes the decision process.

# Portfolio for Sustainability: Native American Behavior Blended with Western Science

# Chapter 4

# The Nuts and Bolts of A Sustainability Portfolio

*"Humankind has not woven the web of life.*
*We are but one thread within it.*
*Whatever we do to the web, we do to ourselves.*
*All things are bound together.*
*All things connect."*

~ Chief Seattle (1854) ~

The western world has to adopt a resource sustainability portfolio that deals with human and environmental vulnerability cycles to disturbances, e.g., human and natural, and climate change. A sustainability portfolio cannot just adopt one element of this toolkit and expect to become a sustainable practitioner. Utilizing the complete portfolio allows humans to adapt to social, economic and environmental boom and bust cycles (Fig. 4.1; see section 10.5). Boom and bust cycles are normal even though they occur at unpredictable times. When trade-off decisions are not made for boom and bust cycles,

- Decisions are made reactively to a problem and not proactively to develop a strategy to deal with a potential problem;
- Decisions are typically formulated by special interest groups so the priority to develop consensus is low;
- Decisions are made to address what appears to be the immediate problem and not other attributes or repercussions of the problem that make people more vulnerable to future situations, and
- Problem identification is egocentric and not at the community level, e.g., what is my problem and how will I be impacted by it.

Western world societies have been fixated on resolving today's problem but without understanding that, as frequently mentioned by Native Americans, "*the mountain will be here after we are gone*". When long-term end-points are identified using short-term knowledge, the problem frequently will be misdiagnosed. It

**Fig. 4.1** An image of western world scientists collaborating with Native Americans to make decisions on global natural resource problems.
Illustrator credit: Ryan Rosendal.

is important to recognize that environmental and social volatility and changing societal values (see section 10.5) make it difficult to identify and plan policy using scientifically identified final end-points. Too many unpredictable events can impair reaching these final end-points. This makes it difficult to adapt to cycles in nature and society since leadership has written the road map and most people find it difficult to take a detour.

Many problems result from boom and bust cycles that occur in society and in our environment (see section 10.5). So decisions to be sustainable need to be adaptive and change goals as new information emerges. Since most of our environmental and social problems exist for decades, a new approach to practice essential leadership is needed [64]. One written plan or solution is not going to work after a couple of years. We need to recognize that the "problem" may still be around even after we are no longer dealing with it. This does not mean that you do not make decisions but you have to be more creative and adaptive in how you follow the road map that you drew.

The western European approach to deal with complex environmental and social problems is frequently a *discipline-based approach or a bandage to cover up the problem.* Today, most Americans are too "temporally" and "spatially" isolated from their environments and the products they consume. So they are less able to know when a solution to a tricky problem will make a difference or when it is a "bandage" and the problem will erupt in the future.

This situation is commonly found in forests today. Over the last 20 years, forests have become the "common property" of global societies. The role of forests in impacting global climates, and their high biodiversity make industrialized societies want to determine what activities or uses should occur in them. Therefore the western world policy focus is on conservation and the continued delivery of ecosystem services from forests [29]. Historically, there has been less in the native people who live and survive in these same forests [29]. Making the trade-off decisions in global forests is tricky because the western world does not want to compromise the delivery of conservation values from these forests. This down-river approach to resource management and conservation does not address the needs of native peoples, the up-river people. Since western world consumers do not have inter-generational connections to forests located at distant parts of the world, it is difficult to engage them on the real trade-offs that need to be made.

Ecosystem and adaptive management are concepts developed by the western world in an attempt to better link society to nature. The concepts are important because they begin to humanize nature management. But the tools development has been elusive for most western world citizens. This is not the same situation for native peoples. Hundreds of years ago, native peoples practiced a form of ecosystem and adaptive management. It is worth having a closer look at what worked for native people over several century time scales. It worked during nature's boom and bust cycles when scarcity was real.

Native people did not call their practices using the same terms as invented by western scientists today. The term is not so important as the practices themselves are. Before we start thinking about the elements of our portfolio, we want to describe what practicing nature management is to native peoples. We want to examine how the ecosystem paradigm developed by the western scientific community is similar to traditional practices. We also want to make the reader think about how diverse native peoples are culturally and there is not one model of what a traditional practitioner is.

The western world has spent over 20 years developing the attributes of ecosystem management. The principles of ecosystem management are very similar no matter whom you talk to around the world. This uniformity is a problem if one wants to be a sustainable practitioner. Sustainable practices cannot be uniformly defined as a common set of tools used by everyone. A common set of tools will not deal with local variations in the lands productive capacity or the disturbances that occur in different space and time situations. Practices or tools need to be made locally relevant to a group of people to be sustainable. It is critical that we do not form one definition of what it means to be a native practitioner. Some native people are less willing to accept western world ideas while others are very open to adopting technology from the west. Each has designed different practices that are suitable for their cultural norms. Even when native people adopt western technologies, they want to maintain their culture. They do not want to move into a large city and become part of the "melting pot" of the intrusive society that can be found in these places.

## 4.1 Practicing Indigenous Cultures and Traditions

Our story from Indonesia and earlier Native American stories highlight

- How indigenous communities established complex landscape management plans that designated the intensity of land use that would be allowed in each part of the forest. Some areas of the forest were set aside as sacred areas, i.e., no land-use due to cultural beliefs so any form of land-use would be prohibited.
- Nature has always been an important part of indigenous communities and nature defines communities cultural beliefs, and
- Indigenous communities do not follow one model of landscape management but always include their local cultural beliefs to adapt to their environment (see Fig. 4.2). There is not a one-to-one correlation between the degree of traditional nature knowledge held by a community and whether the community is ranked as being sustainable.

Each community will have a different place in our figure (Fig. 4.2). Each community will have a different balance between nature and society/technology and still be sustainable practitioner. There are multiple models of sustainability. The key point is that as a community moves further away from nature, it will find it more difficult to be sustainable when consuming natural resources.

In Figure 4.2, we introduce the configuration of three different societies that are influenced by natural endowment, traditional/local science, and modern/global science; this descriptive model evolved from our stories already written in Vogt et al. [1] and what is included in this book. The Inner Baduy and uncontacted Amazon tribes live in the area where natural endowment is available to them without any significant disturbance from other interests/technology. These two societies implement their traditional knowledge that may include both their adaptation and mitigation strategies. Modern/global sciences have less influence or do not have any influence on these communities. Although most anthropologists believe that those two communities interact to some degree with other tribes.

The above categorizations of the Dieng Plateau are being fully implemented by customary societies (society with White Dot in Fig. 4.2), but they have less

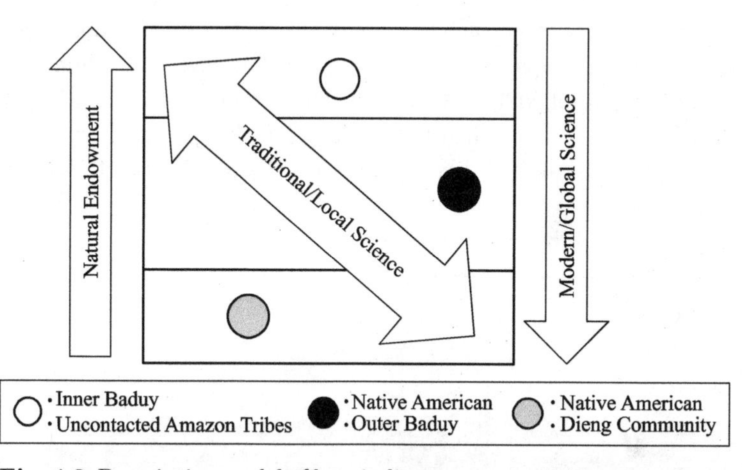


**Fig. 4.2** Descriptive model of how indigenous communities vary along several gradients for how much traditional and local science informs how they manage their natural environments.

impact on societal survival compared to the Black and Gray Dots. Halimun National Park currently is facing a serious threat from local people who do not live with traditional ethic societies but have more modern and economical points of view on resource consumption. Illegal logging at Halimun is not done by people who live adjacent to the area, but by people who understand the value of the timber in local and national markets. Indigenous peoples who live in the buffer zone of Halimun only harvest timber to support their subsistence livelihood.

Natural endowment is supposedly abundant for these indigenous peoples. Dayak Punan (Malaysian Borneo) and Dayak Ahe (Indonesian Borneo) follow conservation practices in their daily life. Preservation, utilization, and protection are part of the life of Dayak People.

They have cultural traditions on how to cultivate and to grow rice to support their life. In fact, rice is the soul of Dayak (Dayak Ahe that live in the Western part of Indonesian Borneo, personal communication, Asep Suntana 2012):

> "Without rice, Dayak will be confused and drunk. If Dayak is drunk, anything can happen uncontrollably"

The elder of Dayak Ahe has been furious for years after he and his people learned that growing rice is going to become more difficult due to increasing number of plant diseases and pests. The Dayak people are also less able to acquire clean water for both their daily life needs and/or to water their crops. The quality of the water and the sedimentation of river reaches are preventing Dayak Ahe from implementing their traditional practices to cultivate rice and other plant for their basic needs.

The landscape changes occur at such a large scale that they can no longer control its impacts on their crops (see Fig. 4.3 right). The river is polluted by chemicals that are being applied far away from the village. The pollution is occurring in the upper part of watershed but is changing the quality of water in the lower part where the Dayak Ahe live. Therefore, the quality of life for the Dayaks has changed.

**Fig. 4.3** Left: Potato plantation at Dieng Plâteau, Central Java, Indonesia (Photo: Asep Suntana, 2009); Right: Palm oil and rice being grown in the transmigration area in West Kalimantan, Indonesia. The areas are managed by migrants from Java island. Photo courtesy: Asep Suntana 2012.

How the practice of traditions and cultures varies in one region is described for the Halimun ecosystem area in Indonesia. Some of the communities have avoided any cultural practices that include western influences while others are very comfortable with integrating western technologies into their everyday life and decision processes. So there is not one model of what a traditional practitioner should behave like. We need to avoid collapsing all traditional practitioners under one umbrella. The key point is that all of these variations of traditional practitioners characterize how communities have adapted to their history and lands potential to deliver ecosystem services. The communities living in the Halimun ecosystem are excellent examples of the diversity of forms that culture can encompass. This will be briefly described next and highlight how different people settled in this region and developed their cultural practices.

### 4.1.1 Cultural Forest Practices in the Halimun Ecosystem Area, Indonesia

The Halimun ecosystem area is located in the upland region of the western part of the Island of Java, Indonesia. Administratively, this area is spread in three districts of two different provinces, namely, West Java and Banten provinces. Forests in Halimun ecosystem area are known to contain the remaining mountain forest ecosystems in Java. The Halimun area consists of three main types of ecosystems, namely, lowland rain forests, sub-montane forests, and montane forests. The flora found in these forests is highly diverse with species record counts reporting 250 species of orchids, 12 species of bamboo, 13 species of rattan, and various types of timber trees.

The Halimun ecosystem area is also the last habitat for various wildlife species and rare endemic animals. In addition, this area has important hydrological function for three provinces: Banten, West Java, and Jakarta (the capital of Indonesia). Due to the critical ecological services of the area, the government established the Halimun National Park. At the same time, certain parts of the Halimun ecosystem area were officially allocated as mining areas and agricultural estates. With the licenses issued by the central government, both state-owned and private companies manage mining and agricultural areas in this Park. The upland agricultural area has been managed since the colonial period.

However, certain parts of the Halimun ecosystem areas are homelands for two different customary communities, namely, the Kasepuhan community and the Baduy community. They live on the same lands that were officially allocated for mining operations and agricultural estates.

The customary communities have inhabited this area for many generations. According to Kusnaka Adimihardja [65] the history of Kasepuhan communities has some connections with the fall of the last Sundanese Hindu Kingdom, i.e., Padjadjaran Kingdom, in the year 1579 CE. Around eight hundred members of the Padjadjaran Kingdom (mostly members of various hierarchies within Special Forces of the Kingdom) managed to escape and hid in several remote areas in upland Banten. These people formed several social groups that are collectively known as Kasepuhan.

Some members of the Padjadjaran Kingdoms joined the Baduy community, a customary community that lives in upland Banten [65]. This Baduy community has been known for their efforts to maintain traditional practices for the purpose of "maintaining the purity of the earth" [66]. They still practice their traditions today.

The Kasepuhan community, which consists of eleven sub-groups, implements a customary system of community-based forest management (CBFM) that is combined with a customary system of forest tenure. Many areas within the customary territories established by the Kasepuhan Community are communal lands.

According to one sub-group of the Kasepuhan community, the areas that are designated as communal lands are the customary protected forests (Sundanese: *Leuwung tutupan*), customary reserved forests (Sundanese: *Leuwung titipan*), and managed forests (Sundanese: *Leuwung garapan*). The first communal area—the protected forests (*Leuwung tutupan*)—is designated as a protected area in order to maintain the ecological, socio-cultural and spiritual functions of the forest. The second communal area (*Leuwung titipan*) is reserved forest lands that can be used during certain times (through specific access mechanisms) to meet the demand for certain basic needs of the community, including food, medicines and housing materials. The third communal area is categorized as forests that can be utilized or converted to other uses (*Leuwung garapan*). This third area consists of forest land that can be "opened" by Kasepuhan Community members and converted into dry agricultural land or for use as mixed gardens [67].

Other sub-groups of the Kasepuhan Community divide customary lands into three forms of land-use. The first form is Old Growth Forest, known as *Leuweung kolot/Leuweung geledegan.* It is a dense forest, e.g., forest with lots of trees and various forms of wildlife. Some members of these communities categorized the Halimun National Park as *Leuweung kolot.* The second form of land-use is "production forests" (*Leuweung sampalan/Leuweung bukaan*). This land-use category allows the harvesting of timber, developing lands for agriculture, the collection of wood fuels, and animal grazing. This forest is usually located in the area that is close to a village. The third land-use category is the sacred forest (*Leuweung titipan).* Gunung Ciawitali and Gunung Girang Cibareno are *Leuweung Titipan...* "Nobody is allowed to utilize any forest products from this forest without having permission from Customary Leader(s)" [1].

The two customary communities have managed the forests of the Halimun ecosystem area based on a set of traditional knowledge rules to ensure the sustainability of both the forests and the livelihood of the communities. By applying custom-based forest management systems, these two customary communities have protected forest resources of the Halimun ecosystem area.

However, since the 1970s, these customary communities have faced restrictions in applying their customary land-use systems since the area was designated as a "state forest zone". In the case of the Kasepuhan community, which consists of nine different sub-groups, their customary territories cover almost the entire Halimun ecosystem area. The establishment of the Halimun National Park in 1993 changed all the rules for the Kasepuhan community. In 2003, the park was further enlarged. These changes have severely restricted the practices of the Kasepuhan community and their ability to implement their traditions on their customary lands [67].

Besides these two customary communities, the Halimun ecosystem area is also home for other local communities. Some groups migrated into the area during the colonial era to work in the agricultural plantations that were managed by a colonial state company. Others arrived more recently to work in the mines located in the Park area. Over the years, these local communities have managed state lands as rice fields and complex agro-forest systems. The area where they

---

1 Sesepuh Girang, personal communication, Mia Siscawati 2012.

live and cultivate is now part of the Halimun National Park. This threatens the continued cultural practices by these indigenous communities because of many new competing demands from these forests. These communities now need to adapt to whole new set of constraints that impede their uses of customary resources and their ability to implement customary land-use practices. This is a very similar story to what the Native Americans faced in North America upon the arrival of the European colonialists (see section 2).

The differences between management decisions practiced by indigenous communities and colonialists are still detected today. Lu et al. [68] reported that in northern Ecuadorian Amazon that

> "five indigenous populations studied affected the forest to a much lesser degree than colonists. From 1986 to 2002, indigenous areas exhibited substantially lower rates of deforestation and there was a higher proportion of the landscape covered by primary forest. Furthermore, several measures indicated colonist lands exhibited greater forest fragmentation. The conservation implications of these findings speak to the value of indigenous lands in maintaining forest cover..."

The story just told for the Halimun National Ecosystem area highlights that there exists a diverse set of nature practices. It means that there are a diversity of native people's practices and not one model that the western world should follow. In section 4.1.2, we will follow our portfolio that native peoples can behave like a Native American following sustainability practices. Our sustainability practices focus more on the elements of practices that we should follow if we want to make our practices more sustainable.

### 4.1.2 Essential Practices of A Sustainable Portfolio as Summarized by John D Tovey[1]

Using the old economic models of maximization and compartmentalization *and* the more recent notions of complexity and human coupled systems can be very abstract, laborious in decision making. It invariably requires an external expert to direct or mediate the process. There is something truly beautiful about traditional natural resource management. The only term that describes it best is the Chinese Taoist philosopher Laozi's description of "wu wei". Wu wei can be roughly translated into "non-doing" or "working without working". It is state of doing something without doing it. This is how we envision the vast majority of natural resource management and even human development such as the development of agriculture, development of culture, development of arts, and de-

1 Cayuse, Joseph Band Nez Perce.

velopment of complex governmental systems. It was just done without doing it. There were no academic experts, complex computer models, or "key indicators for success". It was just *done.*

But how can work be done without doing it? Three things are needed for this art of wu wei to integrate the essential and traditional ecological knowledge into any sustainable portfolio. These are described as follows.

**First action:**
The first is to simply trust your gut. Simply, does it *feel* right? While this may sound simple, the ability to trust one's self either individually or collectively requires a lot of knowledge and experience. We seem to partition some with knowledge and some with experience when in fact these two elements need to be brought together to the same table as individuals as well as philosophically into the same person. I can't count the number of architects I have met that couldn't properly hold a hammer, or professional nutritionists that couldn't boil an egg, or accountants that are in debt. Knowledge and experience must be rejoined and equally valued.

Then it must be respected. In preparation for this book I discussed with my grandmother about the nature of leadership within the culture historically. Most text books will list famous tribal leaders: Chief Joseph, Chief Sealth, Sitting Bull, Geronimo, etc. But, she said, most of these leaders were the war or diplomatic leaders because that was the situation they were in. Any other day, Sitting Bull could have easily been subordinate to the Chief on camp movement, Chief of food preparation, Chief of the hunting party, or any other specific task that required a leader of knowledge and experience. Moreover these other leaders interacted seamlessly and on an as needed basis and only as far as their knowledge and experience allowed. The Chief of the hunt didn't come home from a successful hunt and exclaim he (or she) was now also Chief of village construction because his knowledge and experience weren't in those areas. Hierarchy was a complex relationship depending on the task at hand, the knowledge a person had and their personal experiences, which invariably gave rise to the veneration of elders. Today the respect of elders in western cultures can easily be regarded as polite deference, but historically, it was the repository of knowledge and experience. *The elders were the Google Search Engines of pre-Columbian knowledge.*

**Second action:**
The second action that is needed to integrate traditional ecological knowledge into contemporary natural resource management is changing the paradigm of contemporary notions of space and time from a linear notion to a cyclical notion.

Linda Smith, an indigenous researcher from New Zealand and the author of *Decolonizing Methodologies*, points out fundamental elements that are rooted in western philosophical ideas about space and time [69]. Definitions of space allow us to place positional attributes on objects: inside, outside, top, bottom, etc., but they also allow us to describe space by function such as public/private, city/rural/wilderness, home/work/play. These labels are lenses we use in western thought to make sense of the world even if that world is inhabited by people that don't use these conceptualizations. If the idea of the individual and society are very different depending on the circumstances, then public and private space must also have a very different nature. In western tradition the private is based on the activities of the individual and perhaps to the family, but even home construction in the last 100 years has seen the size of common spaces such as living rooms reduced relative to the increase of personal spaces such as bedrooms and bathrooms which emphasize the individual. Pit house construction of the pre-contact Columbia Plateau tribes was generally a pit of about 0.9–1.2 m deep and 7.6 to 12.2 m in diameter. Each dwelling could house up to around 30 or 40 people [70] meaning even the most private activities of life (such as procreation) were a decidedly public affair.

Smith points out three concepts of spatial vocabulary that markedly change the dynamic of a conversation with many indigenous peoples [69]. These three terms are "The Line", "The Center", and "The Outside". Throughout my education and professional background in Landscape Architecture and Urban Design, I've always been made acutely aware of the power of the line. As a designer, one had to be aware of the permanence of placing a hedge, a building outline, or even a road. The hedge may last 5 to 20 years; a building, perhaps 20 to 100 years or so; and a right-of-way could be there forever so long as civilization as we know it preserves it. Geopolitical boundaries are more abstract than right-of-ways, but can have an enormous impact on what resides on either side of that boundary, be they people, plants, animals, stones, rules, etc. And as Smith points out "The Line" is used to describe, demarcate and ultimately to claim lands [69]. This Line was rarely in a fashion that was recognizable to the residents that lived there and certainly did not demarcate ecological functions accurately on either side of this line.

The other two terms ("The Center" and "The Outside") are relational. I like "The Center", for me, and I believe for Smith, can be a mental and spatial space. As a mental space "The Center" is an ethnocentric term that serves as a reference point. It is where knowledge is stored, where political discourse occurs and where power is held. As a spatial reference, it is where we come from and what we come from. "The Outside" is what is not within "The Line" that is around "The Center" and inherently denotes an "otherness". This "otherness" is to be studied out of curiosity as the most benign interaction, or is something that is to be feared, overpowered, conquered, or destroyed as being the most egregious. And since this "otherness" is often not understood, Smith argues, it can be deemed to be "empty" and thus claimed. An example of this was the idea that Native Americans weren't farming the land in a way that was immediately recognizable by European settlers and since "the Indians weren't using the land, someone might as well utilize it, and that might as well be us."

Smith [69] notes how the western understanding of time had a dramatic affect on the interpretation of indigenous research. During the rise of the Industrial Revolution in the 18th and 19th centuries we developed ever more specific time. It became a way for people to organize their activities in increasingly complex social interactions. Before a farmer would start working at "day break", go to lunch "at midday", and go to bed "at night". During the Industrial Revolution, the systems (factories) didn't start when the individuals decided to get out of bed or with the changing motions of the Sun. All the workers needed to be in specific places at a specific time to man the complex machines, and they worked for a specific period of time. This extended to other aspects of people's lives such as church, recreation, and eating, which affected the daily schedule of those at home making meals. This activity had to be scheduled around one's work, which was a higher order of responsibility. When western researchers observed indigenous peoples relaxing and socializing in the middle of the day, finishing work early in the day, or starting work late in the evening, they were considered lazy and indolent because it did not conform to western notions of productivity.

Smith, for the most part leaves the discussion of time here, however more could be added to the discussion. She could have included the connection of time to the recording of history and more about the differences in the notion of time. This is because time in the western sense is directional and positional with a before and after, while many indigenous concepts of time are cyclical or at the very least less temporally positional. My language of the Cayuse and Nez Perce is known as Nimiipuutimt and literally translated as "The People's Language". (Cayuse as a language had all but died out before European contact.) Nimiipuutimt is a remarkably baroque language and is not easy to learn and as such I know very little of the language other than words and phrases. What I do know of the language includes a curious arrangement of time. The language has a present, past and future tenses, but the concept of "present" has a temporal space of a day or more depending on the context, where a contemporary conceptualization of "the present" can be on the order or minutes or less. Future tense was rarely used because tomorrow was not expected to be all that different from today just as next year wasn't expected to be all that different than the last; and there were technically two past tenses which were broken into actions that have happened within our lifetimes and memories, and a "Legend Time" tense that was frequently used in storytelling.

Because of other western phenomena such as detailed recording of history it becomes easier to compare "us now" versus "us then" and this comparison leads to a constant obsession with betterment. This is opposed to a more cyclical approach to time that is closely in tune to the natural environment where the people lived. People finished work early in the day because there was no more work to be done or simply it was too hot in the afternoon to work. People could have started working late because that's when the fish are running or the game come out of the forest to feed in the meadows. Most importantly, there was less of comparison to "us then", which lessened a sense that there was something higher to attain. They just *are*, doing what they are doing, today, yesterday and tomorrow, whereas in western culture there is always a higher office, new fashions, bigger houses as status symbols, and other accumulations of capital. Certainly, this is a comparison of rates of change rather than stating that indigenous people don't "aspire". If that were the case no technological advancements among indigenous people would have taken place at all but the progression from stones, to spears, to atlatls, to bow and arrows is clear. Moreover, I would argue that the aspiration that is present within the indigenous communities was based in the notion of society and community being the basic social building block rather than the individual which is to be discussed next.

**Third action:**
The third is to change the conceptualizations of the *individual and society*. Linda Smith finds the roots of this difference in classical Greek philosophy which diverged from naturalistic to humanistic explanations of the world [69]. Here is where we find the beginnings of the dichotomy of the understanding of the world where humans are separated from the natural world because of the gifts of language, culture, and reason. In research, we describe things and the idea that humans are separated from nature; the mind is separated from the body, etc. Understandably, there is a purpose to this separation. It was easier to study the natural world without considering human behavior, and it is easier to understand the mind without considering the body. This is reinforced by professional boundaries and the fact that it is now easy (perhaps even required) for a person to become hyper specialized in one aspect of the world to the detriment of their understanding of that knowledge in a greater context.

These constructed dichotomies extend to the human realm as well with the conceptualization of the individual and the group. Smith notes that in Western thought, the individual is the basic social unit. The emphasis on the individual is reflected in the understanding of social and cultural development, most western religions and their individualistic relationships with God, and the clear emphasis of the individual through accumulation of capital. Most (Smith would contend all) western research and descriptions of indigenous people are through this lens. While Smith implicitly assigns morality to this, I tend not to because it is just the way we try to make the world more understandable. I do however ascribe a moral attribute to it when it's not at least recognized that the very lens is just that, a lens by which we try to understand. Or, that the lens and perspective of those that are studied do not necessarily use the same lens.

A clear example is the differences between the western and many tribal penal codes. While western penal codes prosecute, incarcerate, and attempt to reform the individual because the *individual* is damaged or in some way needs to be made better, many tribal protocols, from my experience, begin with the notion that the *community* has been damaged and the community needs to be made better. This is not always an easy thing to do and it is conceivably easy to deflect blame from the individual to the society in an attempt at absolving the individual of any responsibility. But this argument is the very western imposition of a dichotomy of the individual and community. If the individual is *inseparable* from community, and the transgressor is the focus of the healing of the community, there is no loss of focus of who is responsible. Rather the transgressor is made well in the midst of community instead in the vacuum of themselves as an individual. A simple metaphor that works for me is if your left hand is cut, you would use your right hand to bandage it, you wouldn't make it bandage itself.

We now extend this notion of separation of self and community to the separation of human activities and natural processes. Most of human history lived with this notion of inseparability between themselves and the natural world. They were always joined; even the overlap between the populating of Western North America and deglaciation and reforestation indicates that the people have been here just as long as the forests have been. Forests are *always* influenced and managed by humans and vice versa [71, 72].

The 2,000 years of western thought that divorced them from nature is at least conceptually over today, but a general understanding of an individual's or community's thinking is nowhere near this reunification. This will take time, and it will take something that academia is loath to provide. This requires the ethics, morality and sense of cultural and personal survival for the final reunification to occur. It is one thing for a learned person with great experience to say we should no longer see the lines we draw between cities and suburbs or between the fields and the national forests because they are all connected and necessary for their mutual benefit, and they have the computer models to prove it. But it takes a moral and cultural shift for the individuals to *feel* that connection and recognize that they are not a separate observer or manipulator of that system. This is no easy task.

Traditional traditional ecological and cultural values are inter-generation practices that set the context for essential sustainability practices. Maintaining ones Indianness and following essential sustainability practices are essential. To reach this goal, Native American behavior and practices need to be blended with the western world perspectives. What it takes to humanize sustainable practices will be briefly discussed since this is a critical element of any portfolio of practices.

## 4.2 Humanizing Sustainable Practices

Despite the hundreds of years of domination by western European countries, Native Americans survive and continue to adapt to the new challenges they face. This does not mean that Native Americans have solved all their problems that arose upon the arrival of the western European colonialists. Far from it! They face high poverty and continuing battles to improve and protect their livelihoods, cultures and traditions. But they have adapted and still practice a respectful and unique form of system-based approach to nature and resource consumption (see section 5). Many attempts have been made to make Native Americans use tools developed by the western world for them to interconnect with nature. This strategy is to move Native Americans away from their traditions.

Native Americans have avoided following this pathway even though it would have made their life easier. It would have required them to become civilized in the mode of the western world. They continue to search for solutions that are culturally acceptable to them for the regional problems they face. Native Americans are not only working on problems found on the reservations. They also collaborate on natural resource problems found on their customary lands even though they do not own these lands—they are owned by non-tribal communities today (see section 10). Native Americans work on problems even if they have no economic return from these collaborations.

An important question to ask is "How were Native Americans able to survive all the threats to their culture and way of life and to emerge as important contributors to the current regional economies while maintaining their diverse cultures?" Non-tribal battles against Native Americans cultures provided tribes the glue to bond and to fight to preserve their culture when facing foreign intrusions. A parallel could be drawn to the climate issues that contemporary cultures have to address because they are so vulnerable to its impacts. This is the story of our book. It provides the clues for building our portfolio of practices and behavior, i.e., humanizing sustainability practices. Understanding this answer provides the clues to how a people can make sustainable choices while adapting to a changing environment. This also recognizes that people cannot be removed from nature and nature is not better off if there were no people living in it.

Facing tremendous pressures to conform to western world ideals and cultural norms, Native Americans could have made a comment similar to Dr. Peter

Venkman in the Ghostbusters movie. They could tell the western world to back off and leave us alone. A Native American would say "Back off, man. I'm an American Indian". However, a smart survival tactic does not result from telling someone to leave you alone. They usually do not leave you alone and will make a special effort to force you to change. American Indians have survived because of having an inner core of strength and stability that their culture and community provide them. Since Native Americans think about the group or village and do not focus on the individual, they are better at making ecosystem-based decisions and responding to disturbances that cycle through human landscapes.

Two key elements of humanizing sustainable practices are: people have to have cultural core is local or regionally placed; and Inter-generational nature knowledge sets the context for community decisions. Following these two elements, a tribal community will less likely make important decisions based on what is the current global fad.

This contrasts the western world view that appears to be driven by fads or public polls. The western world values the individual. The individual is more susceptible to making decisions based on current fads, having what your neighbor has or responding to new media communications that are similar to your values. Only in the western world can we connect the length of a woman's skirt to the stock market, e.g., when the economy is poor, a woman's skirt is longer (//moneymorning.com/2012/08/20/what-skirt-lengths-tell-you-about-stock-market/).

> "Dr. John Casti, Cofounder of the X-Center in Vienna and author of Mood Matters: From Rising Skirt Lengths to the Collapse of World Powers puts it this way, observing 'that the rise and fall of great civilizations are biased by the attitudes a society holds to the future.'
>
> As a leading proponent of the science of socionomics, Casti observes that when social mood is positive and optimistic, hemlines tend to be shorter. And, that the reverse is true when things are dour or the social mood is uncertain because they reflect the mass psychology."

This does not mean that during bad economic times, women should not go out and buy longer length skirts to wear. Wearing a long skirt is not going to reverse the economy or keep a bad economic situation from getting worse. These two bits of fact are correlated to one another but are not causally connected to one another. Culture does not cause fads to emerge since culture is dictated by hundreds of years of traditions.

A people's culture or folklore is the thread that connects current societies to their past history and values. It sustains them when foreign intrusions occur into their culture. A people's culture is the glue that keeps them bonded and working towards common goals, and practices.

It is pretty obvious that the first thing a conquering people want to obliterate is a people's culture. If you can eliminate someone's culture, it is easier to control them and therefore their resources. A conquering group will want to eliminate anything that binds a group of people together. It was a common practice of rulers to adopt children of the defeated rulers. These children were retrained and educated to be loyal to the new ruler. It was an effective indoctrination approach and made it less likely that these children would attack the conquerors in the future. This also helped to pacify defeated subjects who knew that a few children survived and perhaps could retake the throne in the future.

Since Western Europe stories and culture have dominated the decision process for the last 500 years in many parts of the world, few sustainability stories were written during this time. This is why we need to explore native people's stories.

Contemporary forest management practiced by tribes is frequently acknowledged to be better than what is implemented by public agencies overseeing public lands [73]. Tribes have to live with the resulting consequences of any decisions that they make while the public management agencies do not. If a bad decision was made by a public agency and trees die, there usually is not a public outcry to fire or eliminate the agency. However, if this happened on tribal lands, the land manager would hear about it that day and would have to account for the decision made.

Starting about 20 years ago, the western world began to cognize the concepts that were integral to Native Americans. They called these *ideas ecosystem and adaptive management.* These terms will be discussed next. It is important to recognize that even though the definition of ecosystem management includes people [1], it is really not part of the management practice. It is important to recognize these terms before we discuss the portfolio of sustainable practices in more detail.

## 4.3 Western World Ecosystem and Adaptive Management

*"Dr. Peter Venkman: Back off, man. I'm a scientist."*

Ghostbusters (1984) movie—three unemployed parapsychology professors who run a business exterminating ghosts

~ www.imdb.com/title/tt0087332/quotes ~

If practicing holistic management is the goal, we need to accept the fact that nature is not virgin and people are part of nature. In contrast to Native Americans, western world peoples believe that nature is untouched by the hand of man, and nature without a person or human activity is more sustainable.

Accepting people as part of nature is not a betrayal of one's conservation values as believed by many western world ecologists. Yes, we have stories where humans altered ecosystems until they became less resilient and societies subsequently collapsed [17]. But, most examples are of locally-based or indigenous societies who balance their collection of natural resources so that the environment and society remain resilient. If they did not maintain resilient environments, they as a community would not be alive today. Indigenous community survival is dependent upon how well they treat nature.

The western world, until quite recently, saw nature as separated from society and where a society's land-use footprint will always be negative towards nature. This view probably is a historical legacy of European colonial over-exploitation of lands and resources. The environmental impacts of resource over-exploitation eventually made U.S. citizens conclude that society and nature should not be connected too closely together. The western world wanted to conserve biodiversity now but did not recognize the role native people had played in managing nature. Today, we recognize that we are rapidly losing many ecosystem types because the western world eliminated native practitioners from doing their job.

Today, the western world conceptually recognizes that ecosystems and societies are inextricably connected to one another. But this insight did not arise in western world mind in any systematic way until the early 1990s. When ecosystem management and adaptive management were conceptually developed, western land management agencies accepted and developed new paradigms for land management. This was an exciting time for the land agencies as management was shifting to a system's based approach.

John Gordon's definition of these terms is very descriptive and articulates well what these terms mean and also how they differ from one another—a fact that is mostly forgotten by decision-makers and scientists to their detriment (see Box 14). Gordon's definitions recognize that resource scarcity is driven by several factors: a scarcity of land and a scarcity of knowledge. When we ignore these scarcity differences, we tend to adopt tools that are inappropriate for a situation.

If you recognize that ecosystem management deals with scarcity of resources from any land, you would have to accept the need to address competitive use options from the same land. You might use economic tools to address the competitive choices but, after reading this book, we hope you will use tools that are not just based on an assessment of costs and benefits.

Since adaptive management is really based on a lack of knowledge, you can design your tools to identify factors that will reduce the risk that you make bad choices. Adaptive management is more difficult to evaluate using only economic tools since the knowledge to define thresholds and end-points is essential information for an economic assessment. This knowledge is mostly lacking for our

**Box 14: Terms Describing Scarcity of Land and Scarcity of Scientific Information**

**Ecosystem management** is attempting to respond to a scarcity of land. This land scarcity means that the exclusive uses of any given land area for one purpose is no longer possible because the production capacities of ecosystems are finite.
**Adaptive management**, managers are responding to scarcity of scientific information and knowledge.

(from Gordon [74])

natural environments. If scarcity of knowledge is real, it will be difficult if not impossible to develop consensus on which tools to use and to accept the results of any assessment [1]. This is where traditional knowledge has an important role to play since it helps society to adapt to change under conditions of knowledge scarcity and "boom and bust" cycles.

Western European models of resource consumption and economic development do not work in an environment where resources are scarce and where competition for resources is keen. The western world model of resource consumption, despite all the scientific discussions and writings, continues to be driven by economic decisions. These models include environmental and societal factors that can be readily applied in existing models. Our lack of knowledge on the environment and the occurrence of unpredictable events suggest that adaptive management, not ecosystem management, should be the tool we need to practice more. Today, the norm is to use ecosystem management to address environmental problems that society faces.

The need to make difficult decisions with uncertain knowledge is common. This is especially problematic when non-Indians want to negotiate with Native Americans to access resources found on reservation lands. Today, non-Indians have to negotiate with Native American tribes since societal demand for scarce resource supplies continues to increase. These scarce resources are found abundantly on many reservation lands (see section 7.4). Water is an especially scarce resource that societies cannot survive without and where current patterns of climate change are altering who owns it (see section 1.3).

Despite all the losses of resources and lands that the Colville Confederated Tribes faced with the building of the Grand Coulee Dam (see section 2.2.4), new demands to access water supplies found on tribal lands continue even today. The increasing demand and competition for water is mostly occurring from the many groups who live down-river from reservations (see section 1.3.1). When ecosystem management tools are used to decide how to equitably distribute water, it should address scarcity of water and how much water can be distributed without jeopardizing future water supplies in this region. However, in 2008 it was reported in the Seattle Times that Lake Roosevelt drawdown was approved by the state of Washington and what was agreed on [75].

> "As part of that bill, the state proposed drawing down Lake Roosevelt, behind...of that water will remain in the river for fish, and one-third will be used for new municipal and industrial water rights along the Columbia. The rest will provide surface irrigation for 10,000 acres [4,047 ha] of crops east of Moses Lake, where farmers have been relying on well water from the declining Odessa Aquifer, and to provide a more stable water supply for irrigators whose water rights are interrupted in drought years... "This 20-year war, where people's outlook was, 'I am going to win and the other side will lose,' everybody loses," Manning said. "That's what water has been about, at least in Washington, certainly in Eastern Washington." ...The state reached agreement with the Confederated Tribes of the Colville Reservation and the Spokane Tribe of Indians, whose reservations border the lake, on payment for tribal costs of the drawdown..."

This agreement made for Lake Roosevelt appears to be an equitable distribution of water supplies derived from tribal lands and that tribes would be compensated for their losses. As a rationale scientist, one does have to question how much knowledge was used in the decision process. It is not clear whether the allocation of a third of the water for fish, a third for new municipal and industrial water rights along the Columbia and the remaining a third for irrigation sufficiently evaluated risks to maintain ecosystem services from the Columbia River.

This appears to be an agreement that allocated equal portions of water for the competing demands. What if there had been another powerful stakeholder group, would the water have been distributed by fourths? If we use Gordon's definitions of terms, the Lake Roosevelt agreement was treated as an ecosystem management problem, i.e., how to allocate a scarce resource. In fact it appears to be an adaptive management problem where a lack of knowledge resulted in a simple process of equally dividing a scarce resource that has strong economic ramifications.

Any decision on water allocation for Lake Roosevelt should use the holistic management approach implemented by tribes. The Colville Confederated Tribe pursued Holistic Management because they saw it as a way to factor in cultural factors in every aspect of tribal life.

## 4.4 Recognize Western World and Indigenous Community Differences in How Humanize Sustainability

There is still considerable lack of understanding of tribal practices and cultures. A western world lens is used on tribes. The story below demonstrates the lack of understanding of local tribal knowledge and its relevance for managing resource landscapes (Box 15).

Some differences between Native Americans and western world towards resources and their decision-processes are summarized in Table 4.1. A western educated person has to acknowledge that these differences exist and they modify one's behavior. These differences form the basis of our Native American portfolio that we introduce in this book.

The reader has to recognize that there is no one Native American attribute to mimic to behave like an Indian. Native Americans have a diversity of behavioral attributes that varies depending on where a tribe lives. The reader also has to recognize that there are core cultural roots, values and behavior that glue tribes together even if a specific cultural attribute varies among the tribes. Some of the comparisons we make will not be found within a particular tribe or a tribal member. The fundamentally different approaches between a Native American and a western world person are summarized in Table 4.1. These are the elements of our portfolio that are more extensively covered for the reader in our book.

Our portfolio does provide a road map of practices for anyone that wants to humanize natural resource decisions. It allows an individual to think about how to change their behavior to integrate culture into their thinking, behavior and decision-process. It will help a person to connect to nature without becoming a false parody of the real thing or only making cosmetic changes (see section 5.2).

Table 4.1 summarizes our thoughts and explains why we include some of the elements that we did in our portfolio. We suggest that the western world perspective is rapidly evolving and beginning to move towards the Native American approach. The western world has not reached this goal of becoming and behaving like an Indian. So we apologize to those who have successfully made this transition by having to read our stories.

**Box 15: WSU/Washington State Forest Owners Field Day by Rodney**

In Washington State private land owners control 5.8 million acres of non-industrial forest land. Several times throughout the year, Washington State University Extension and several Washington State Agencies, including the Department of Natural Resources hosts a Forest Owners Field Day for private forest land owners. These Field Days provide educational workshops with information that private forest land owners can use to manage their own forests. Courses include classes and activities in forest health, wildlife habitat, soils, fire protection and special forest products.

While I was employed at DNR, I was asked to conduct presentations at the Forest Owners Field Day in tribal-state relations, tribal government and cultural resources. During these presentations I was asked questions from everything about Casino's to tribal councils, and tribal enrollment. Many landowners never considered that the land they now own was once inhabited by Native Americans. While most of the participants held tribal affairs and cultural resources in high regard, there were some who did not.

Some private land owners didn't like the fact that tribal representatives could make comments on the Forest Practices Review System utilized by Washington State to process permit applications on private and state forest lands. Many didn't like the tribe's involvement with the Washington's Forest and Fish Law that provides opportunity for tribes to make comments to protect habitat on miles of streams and aquatic lands and public and private forest lands. There were private forest landowners who wanted to maximize their revenues and did not like leaving barriers along stream sides or riparian areas.

These workshops were helpful to bring a greater understanding of Native American use of the landscape and afterwards many private land owners invited tribal gathering on their lands. But it was also obvious that a continuing generation of conflict still exists of people who don't realize that the very landscapes that they own with high biodiversity is the result of the practices of the people who have lived there for thousands of years.

(Confederated Tribes of the Colville)

**Tab. 4.1** Summary of native American and western world people's perspective on nature and problem solving.

| Topic | Native American Perspective | Western World Perspective |
|---|---|---|
| View of nature | Nature has no boundaries or borders | Nature has borders and needs to be separated from humans by artificial structures. |
| | The land is managed in longer time scales, where as contemporary management plans tend to expect immediate results and feedback. | Nature needs to be managed in shorter time scales and for immediate benefits. |
| | Humans are part of nature—there is unity between nature and humans; nature is where humans and spirits interact. There is an emphasis on living in balance and harmony with nature and with the spirit world. | Humans are masters of nature and dominate over nature so there is no need to balance nature with the spirit world. Nature not managed is a Park where people can view and photograph its natural beauty but not touch nature. |
| | Nature is everywhere and interconnected. | Nature is found in a museum, zoo or park and isolated from humans. Nature is not interconnected formally to humans. |
| | Everything in nature has a spirit and should be respected. Nez Perce call spirits *wayakin* while Interior salish would say "sumix". The sacredness and spiritual significance of nature and the earth is important; particularly objects in nature such as mountains, rivers, rocks, stars, and plants as well as many animals, such as eagles, hawks, crow, owls, deer. | Nature follows scientific principles and will behave according to science knowledge and principles. Particular objectives in nature are not important as much as species facing extinction or species endangered due to human land-use activities. |
| Nature managers | Individuals-to-community practice adaptive ecosystem management. Bottom up ecosystem managers. | Land managers do not live in the community. Top down ecosystem managers. Adaptive management is not common and mostly managers respond to problems as they occur. |
| Thinking | Web based and systems view of nature and resources. | Linear thinkers and separate nature into categories that are individually managed. |
| Source of knowledge | Traditional ecological knowledge which includes cultural/spiritual values. Elders and shamans are important sources of knowledge. | Academic knowledge developed by experts and does not include cultural/spiritual values. No elders or shamans contribute knowledge to resource issues at the local level. |
| Science learning process | Participatory, experiential learning, inter-generational stories of nature. | Listening and regurgitating back scientifically accepted facts. Less experiential learning. Tools important to assess resources: Environmental Impact Assessments, certification, GIS, simulations, statistics, scientific research. |

(Continued)

| Topic | Native American Perspective | Western World Perspective |
|---|---|---|
| Resource values | Social, cultural, spiritual values for resources. Resources do not need to be useful to be valued. | Resources are commodities, economic values important. Resources have to be useful for humans. |
| Level of decision making | Local and community level decision making. No special interest groups controlling decisions. Society, communal harmony, kinship and cooperation are emphasized. | Decision-making mostly external to local and community level. Special interest groups common and foreigners making decisions. Little emphasis on communal harmony and cooperation. |
| Belief | There exists a Great Power or Great Mystery and various other gods which are nature-based. | Religions have great power and symbols but these are not nature-based. |
| Ceremonies and rituals | There is an emphasis on self-discipline and performing various tests of physical ordeals. Purification practices, fasting and vision-seeking is common. Rituals, stories, dancing, art, chants, and music (mainly singing and drumming) maintain cultures. | Ceremonies are less relevant except as part of each person's religious beliefs. Vision seeking is not common since science or superstitions explain everything. Culture is expressed in festivals where people dress in traditional costumes from one's family roots, but they are geographically foreign to where the festival takes place. |
| Crime | When a person commits a crime, the community is damaged rather than an individual being a deviant. The resolution is a sense of the individual healing in the midst of a community that also needs to heal. | When a person commits a crime they are removed from the community to heal in a vacuum. This results in the community not be able heal and more difficult for the individual to become part of the community. |

## Coyote Essentials

- Western world science developed ecosystem and adaptive management 20 years ago that is similar to Native American nature practices developed hundreds of years ago.
- Western world science decision process is better able to make choices when dealing with a scarcity of land resources but not with a scarcity of knowledge. Native American practices deal with scarcity of land resources and scarcity of knowledge equally well.
- Tribal elders are the Google Search Engines of pre-Columbian knowledge.
- The culture and environmental location of each community/tribe determines which sustainability practices they find acceptable/developed/value.

# Chapter 5

# Portfolio Element I: How to Connect Society with Nature

*All things share the same breath—the animal, the tree, the man, and the air shares its spirit with all the life it supports.*

*The earth and myself are of one mind.*

*All things are connected. Whatever befalls the earth befalls the children of the earth.*

*We are part of the earth, and it is part of us.*

~ Chief Seattle (Seathl) (1790–1866) ~

## 5.1 Divergent Models of "Wild" Nature and How Different Societies Connect to It

The western world and Native Americans have fundamentally different views of nature and how societies connect to it. These differences are important to know since it dictates the approaches taken by each group to make choices with respect to nature.

The review of Anderson's book *Tending the Wild: Native American Knowledge and the Management of California's Natural Resources* by Sally Fallon Morell [76] summarizes well the contrasting views of nature held by Native Americans and western European settlers when they first arrived in present-day California:

> "The Europeans assumed that they had discovered an untouched wilderness that just happened to resemble a garden, populated by 'primitive' Indian tribes who profited from Nature's bounty simply by hunting and gathering. But in fact, California was not so much a wilderness as a

> true garden, a garden of beauty and abundance because it had been tended for thousands of years by wise guardians. For untold generations, the California Indians had shaped the landscape by pruning, coppicing, cultivating, transplanting, weeding, selecting cultivars—and above all by controlled burning.
>
> Controlled burning served as the main tool for creating California's garden-like landscape... Burning could be used to corral wildlife—masses of grasshoppers moving ahead of controlled burns, for example, provided nutritious and easily gathered morsels for the Indians. Above all, frequent small fires prevented the buildup of brush that could fuel the occasional catastrophic fire. Whereas controlled burning helped to preserve trees and encouraged them to grow, uncontrolled fire could wipe out forests and therefore the food supply...
>
> The Indians saw their role as guardians of Nature, agents for improving Nature's appearance and increasing her abundance; the plants and animals were their relatives, to be supported and cared for, just like human relatives. By contrast, the Europeans viewed Nature as something outside—unpredictable and often dangerous; Nature was there for exploitation or, in the case of naturalists like John Muir, to be left 'pristine' and untouched. Interestingly, modern Indians often use the word 'wilderness' as a negative label for land that humans have not taken care of for a long time, a land where dense understory shrubbery or thickets of young trees block visibility and movement. ... The Indians believed that a hands-off approach to nature—above all the prohibition on controlled burning—promoted wild and rank landscapes that were inhospitable to life. 'The white man sure ruined this country', said James Rust, a Southern Sierra Miwok elder. 'It's turned back to wilderness.' "

Two comparisons highlight how strikingly different the connections to nature are for a western world person and Native Americans. Native Americans value harvesting their cultural foods from nature while the average western Europeans buy their organic foods from a grocery store. Western Europeans do not learn from their parents and grandparents how to and when to collect cultural foods. Even today, Native Americans manage their lands using traditional ecological tools like fire so the land supports the growth of cultural foods, e.g., berries, camas, etc. By contrast, western Europeans do not actively use fire to increase the productivity of their berry bushes but instead grow them in cultivated fields. An exception to this generality is mushrooms. Mushrooms are produced by a fungal symbiont associated with tree roots called mycorrhizas. You need an intact forest habitat for these symbionts to grow and produce their reproductive structures, i.e., the mushroom that humans and many other animals eat. Western Europeans collect mushrooms as a hobby from forests even though several species of mushrooms can be grown artificially. There is a long history of western leaders collecting mushroom to relax, e.g., Lenin. Many of the tasty mushrooms are produced by mycorrhizal fungi, e.g., chanterelles, boletus, etc. that only grow well in nature.

The second comparison is the western world idea of becoming a vegetarian to help mitigate one's climate change footprint. The rationale to become a vegetarian is to reduce your carbon footprint by not eating foods that cycle more carbon gas during its production, e.g., meat from cattle grazing on grass. If you decide not to eat animal meat, however, you have to eat plants to survive. For the western world citizen, eating plants is not a bad practice because they do not recognize plants as having spirits. By contrast, a Native American respects all life forms whether it is plant or animal and do not distinguish between one and the other. Native Americans thank the spirit of the food they eat. They see everything from a plant to a rock and to a person as having a soul. For an American tribal member, eating plants can be just as bad as eating animals. Therefore, they think being a vegetarian is a funny or strange idea.

The idea of a plant with a "soul" is not difficult to defend. Scientific research documents plants communicating with one another. At the University of Exeter, scientists recorded plants releasing a gas when attacked that was detected by other plants (www.bbc.co.uk/news/science-environment-16916474). Several years ago, the idea of "screaming" plants rose onto the conscious of the western world when the ability of plants to communicate was detected. If plants do not have any feelings or communicate with each other, is it okay to eat them? If plants do communicate with one another and transmit information when they are being killed, is it alright to eat them? This potentially complicates anyone's decision on whether to become a vegetarian helps the world by reducing your individual carbon footprint.

Despite the western world wanting to make the right decisions related to the environment, the idea of becoming a vegetarian to reduce one's carbon footprint has not been widely adopted by the western world. In most western world countries, becoming a vegetarian to reduce one's carbon footprint is not a common practice since the number of meat eaters is increasing. The Economist reported that the global consumption of meat increased from 70 tonnes about fifty years ago to 268 tonnes in 2007; individual consumption of meat had more than doubled during this time period [77]. The U.S. was ranked second in the amount of meat consumed per person—∼125 kg per person in 2009 [77]. This trend would suggest that becoming a vegetarian is not high on the priority list of the average western world citizen. Therefore becoming a vegetarian is a fad. People are not eating animal meat because animals have souls. This example says nothing about the value of becoming a vegetarian but that most western societies have not formalized it into their cultural practices. Some cultures do not eat animal meat for religious reasons and out of respect for nature and wildlife, e.g., Buddhism.

The dominant western world view of nature diverges from the dominant view held by Native Americans. Native Americans respect all aspects of nature—not just one part. They collect from nature's bounty without feeling the need to bind or dominate nature. They collect and improve the growth of cultural foods provided by nature but do not arbitrarily destroy nature for the fun of it. Their sense of ownership of nature differs considerably from the western world where nature is isolated from people. The western world view is that nature needs to be controlled. It should be demarcated and confined behind walls, e.g., zoos, where nature can be controlled and looked at for human entertainment or enjoyment. In the western world, nature entertains us, e.g., circuses where animals are trained to jump through fire lit circles or stand on their hind legs.

The historical beliefs held by the dominant religions of Western Europe also espoused the dominium of nature by humans. This contrasts the tribes' belief that humans are simply a part of nature and nature does not need to be dominated. When control of nature is the norm, no unused space is acceptable and every piece of land needs to be used and altered for some human benefits. When Europeans colonialists first arrived in North America, they felt it was empty and undeveloped by the American tribes. These colonialists began to develop the economic potential of the Americas for their own benefit. They had Manifest Destiny on their side so it was their right to dominate nature. By contrast, tribes felt that they had harvested from the land for centuries while keeping the impacts of their footprint small on the land. Tribes did not over-consume natural resources while the European colonialists commonly over-exploited nature for its resources for societal and economic benefits.

It is worthwhile having a brief detour to examine why the western world and Native Americans have such divergent views of nature. These differences impact how each society interconnects with nature and defines the common acceptable practices used to collect nature's capital. If one's view is that nature should be isolated from humans, a nature experience becomes a zoo where humans look but should not touch nature or where over-exploitation of nature's resources is easy to practice. There is no middle ground for how nature and society interconnect. Very un-coyote like. If a group of people do not dominate nature or control it for purely economic gain, it is easier to find the middle ground to balance society to nature. In this situation, nature does not become a zoo experience or where it is seen or experienced in small doses; e.g., an hour or two of paid visitation time.

## 5.2 Western World Model: Nature Bounded by Borders

> *"A wild Indian requires a thousand acres to roam over, while an intelligent man will find a comfortable support for his family on a very small tract... Barbarism is costly, wasteful and extravagant. Intelligence promotes thrift and increases prosperity."*
>
> ~ Adams [47] ~

The western European model of nature is integrally combined with their views of what it means to be civilized (see section 2.1.3). The quote above is from a western world citizen. It states how you interface with nature will define whether you are a barbarian or a civilized person, i.., being civilized means that you farm a small track of land and you are not nomadic. This really means that you put borders around nature and control nature for your own purposes and benefits.

When the European colonialists landed in North America, they were foreigners in a land with abundant resources and no grasp of *nature* or the culture of the people who lived in the Americas. They also did not appreciate what nature in the Americas provided. According to accounts recorded at the Jamestown site, the early English colonialists did not have enough food and did not know how to fish or hunt [12]. They could see all the fish and the animals that lived in the sea and forests but were ignorant and too weak to fish and hunt them. They were fortunate that the indigenous inhabitants of these lands that they called "savages" were willing to provide them food [12]. Despite their need for knowledge and being able to obtain much of their food from the American Indians, the colonialists treated them "with contempt" and would murder Indians "with a nonchalant brutality" [12]. These colonialists were also reported to destroy Indian corn fields even thought they were dependent on these same fields for food [12].

Perhaps because of Manifest Destiny, the settlers did not "naturalize" themselves to their new environment and many died. These settlers did not understand the people and lands they conquered. This naivety translates into many European colonialists making very unsustainable choices. Maier et al. [12] recounts

> "...in one New World settlement after another, the men of Virginia quarreled, stole supplies from the common store for their own consumption or to trade with the Indians for their personal profit, and became at times so obsessed with gold that other, life-sustaining activities ceased."

Today, Botanical Gardens and Zoological Gardens are the main nature outlets for the average western world child and adult. Nature outlets needed to be close to where one lived since people are too busy working to make money and have less free time.

Botanical Gardens were built at least three hundred years before the first zoos were ever built in a city. Italy is given the credit for building the first Botanical Garden in the world. At this time, the Botanical Garden began to reconnect the western world societies to nature as recounted on Padua's website (//whc.unesco.org/en/list/824, accessed 13 October 2012)

> "The Botanical Garden of Padua is the original of all botanical gardens throughout the world, and represents the birth of science, of scientific exchanges, and understanding of the relationship between nature and culture. It has made a profound contribution to the development of many modern scientific disciplines, notably botany, medicine, chemistry, ecology, and pharmacy."

In the past, most people had a very structured and exotic visit to a botanical or zoological garden. People were not going to visit these gardens if they housed the ordinary dandelion, pig or cow. This forced these gardens to cater to their customers' exotic interests.

These gardens have not helped the western world connect to nature. Nature has been mostly been viewed through a couple of lenses—entertainment and or its economic values. These views also demonstrated a controlled and very artificial nature which encourages societies to disconnect from nature. Western world societies connect with exotic and unusual nature experiences.

Despite a nature experience needing to be unusual, it has to also be tame and non-threatening. Therefore animal and plant gardens are very structured, they are not chaotic or volatile like nature can be. Ultimately, they become boring for the average western world citizen.

There are repercussions of the western world disconnecting from nature and forgetting how nature functions. The result is an incomplete understanding of nature and what their human impacts are on nature and its environmental resilience. They also do not recognize how human activities in nature feed back to decrease human resilience. It is very difficult to make sustainable decisions when you do not understand how the parts work together and how they are interconnected. It is easy to make decisions that are decidedly unsustainable.

The western world recognizes that it has a nature dilemma but it has been challenging to reverse of this trend that started several hundred years ago. This continues to be a problem today in western industrialized countries.

Today, books are published on children's "nature deficit" and the inability of nature to compete with a multitude of electronic gadgets, e.g., computers, cell phones and games that children as well as adults can play using these technologies. Nature cannot compete with the sophisticated facsimiles of nature experiences that today's media can offer. Nature is so much more colorful and vibrant when offered on gadgets, and one does not have to deal with the smells or wastes that are part of nature.

The story of how the western world built zoos and how they connect to nature is important to tell. This connection is artificial and controlled. It continues the views held by the Euro-American settlers, i.e., Manifest Destiny (see section 2.1). There is a need to understand how society connects or disconnects from nature if one hopes to humanize sustainable decisions.

### 5.2.1 Nature Needs to Be Controlled

A common western world view of nature, especially in the past, was that nature needed to be controlled and wild animals needed to be kept separate from humans. This meant that nature can be looked at but not touched. Nature was also idealized and revered for its beauty and not for the animals that could bite, kill or maim humans. Humans are very quick to kill or hunt animals that are dangerous to them. This is a repeating cycle that humans play out many times in this world. It has allowed humans to survive and dominate this world. Get rid of all animals that are predators on humans. Humans did not recognize how important wild and dangerous animals are in sustaining nature in its natural and healthy ecosystem condition.

When we don't recognize nature as an interconnected system, it is difficult for a society to decide how to protect, conserve or consume nature products so environments and societies are resilient. A detached understanding of nature converts nature into a zoo where exotic animals can be viewed under very controlled conditions. This is not the "real" nature but an image that humans have fabricated in our minds to represent what we think nature is all about. Nature at its best is frequently equated as being isolated from humans. Humans touching or altering nature is always bad. The best nature is one without humans.

The European Union has a category of forests called the "untouched forests". This only means it was "untouched" by Europeans since non-Europeans have been living in these locations for centuries before the arrival of European colonialists. Europeans like to travel to the western U.S. because of the open space that they find there that is so dissimilar to what Europe ever offered. They did not recognize the role of Native peoples in creating these open spaces that they call untouched. So untouched nature becomes linked to wide open spaces and not the cramped nature found in an industrialized or human community. They continue to have these views today and like to be tourists in dude ranches to experience what they think the old west was all about. Most of these values are not based on the reality on the ground but a rose colored glasses view of what nature could be in an idealized world. They did not recognize, or want to recognize, the role of Native Americans in creating these open spaces. This would have forced them to acknowledge that "civilized" people lived on these lands.

When humans develop their infrastructure to support large cities, they are building what ecologists call "novel" ecosystems. These infrastructures are not designed to adapt to disturbances and are in fact decoupled from nature, e.g., they are isolated infrastructures that are not naturally resilient and need continuous human input of resources to keep operating. As soon as humans cease managing nature, nature reverts back to something other than what humans want from this land. Most large western cities were built and designed to optimize the export of human wastes and the import of human survival resources such as food. There is not much nature in these urban areas. If a green space was not managed, it would quickly restart nature's successional processes to produce an ecosystem capable of surviving under a mix of natural and anthropogenic disturbances.

Today, one hears discussions about blurring the edges of cities to make them resilient to human and/or large-scale natural disturbances. But we have a long way from achieving this vision. Most urban dwellers experience nature in a small confined area around where they live, e.g., grass lawn or a city park. Today's cities bring in aspects of nature such as lawns that stimulate the development of several industries and technologies to maintain grass, keeping it from reverting to its "wild" nature, e.g., lawn mowers, fertilizers, pesticides, hedge clippers, etc. These artificial nature constructs would not survive without continuous human inputs of resources and energy. Most city dwellers have little concept of what existed outside of their artificially constructed borders and the type of nature that flourishes in them.

Today's cities are urban sprawls. Cities are isolated from the natural environment and even from one another. For defensive purposes, past cities were built in a way that they isolate western societies from nature and from one another. If you look at European cities of today, they were built when cities needed to be able to close their gates and to defend themselves against marauding soldiers. European cities experienced frequent warfare. Historical fighting between cities resulted in a matrix landscape in present-day Germany where one city is Protestant while another is Catholic. In the Catholic town you hear "Grüss Gott" in the morning while you hear "Guten Tag" in the Protestant town. This pattern still survives today.

History has recorded many battles where the castles and city fortifications of a town allowed a group of people to defend themselves when the fighting began.

Every sizable town had a castle. Each castle needed to secure its water supply when the gates were closed. Therefore, wells were built inside caste walls to ensure the castle's access to fresh water. Fresh water was critical for human survival (see section 1.3). Only by hiding inside the castle walls, and living off resources collected and stored before you were attacked, were you able to increase your chances of surviving an attack.

The western European model of the structure of the built environment was carried to North America by the European colonialists. Thomas Jefferson preferred the houses of Europe and not the wooden houses of North America. He is quoted as saying in 1781 [30]

> "The inhabitants of Europe, who dwell chiefly in houses of stone or brick, are surely as healthy as those of Virginia. These houses have the advantage too of being warmer in winter and cooler in summer than those of wood; of being cheaper in their first construction, where line convenient, and infinitely more durable... A country whose buildings are of wood, can never increase in its improvements to any considerable degree. There duration is highly estimated at 50 years... Whereas when buildings are of durable materials, every new edifice is an actual and permanent acquisition to the state, adding to its value as well as to its ornament."

How this view of nature was maintained by western European descendants can be seen in the changing ownerships of farmlands in the Midwest U.S. Small family farms originally dotted the landscape with towns located between them. Each farm grew numerous crops, and raised cows, pigs, chickens. Their children attended local schools and participated in town activities. Today, large corporations that are usually headquartered on the East coast U.S. own the farmlands. Owners may never have seen the land and seldom visit the property after its initial purchase. These farms have become huge mechanized businesses that usually only grow one crop; whatever is most profitable. Towns that had relied on the small farmers for their existence have become ghost towns. This vision is a good example of a cultural transformation that caused a huge migration where whole educational systems and cultural knowledge disappeared.

### 5.2.2 Zoo Becomes A Nature Experience

It's very interesting to look at all the zoos/menageries that have been built by civilizations going back to more than 4,000 years. They are found all over the world and owned by all cultures from Europe to Asia. Studying the history of zoos, it's very evident that the first animal menageries were simply a way for a country's ruler to display his/her power and to gain status. The more wealthy citizens of that country followed suit, by building their own personal zoos. Animals were given as tribute from one country to another. There was very little interest in the animal as anything but a possession and a status symbol. Quite often the animal, if it wasn't exotic enough to be very valuable, was mistreated or even slaughtered for the amusement of some local citizens in enclosed arenas. Most times its carcass wasn't even used for food.

Today, western European descendents lament the lack of nature experiences for their children as they grow up. A zoo becomes the closest place where humans interact and "bond" with nature. Today's society views nature through a distant lens that is totally tame and unrealistic—it is a "novel" ecosystem. One benefit of this experience is that you do not have to worry about being checked out as a food item or prey by a hunter or predator. Most Europeans killed or eliminated wild and dangerous animals from their environments a long time ago. Humans do not want to deal with a dangerous animal that looks at humans as a nourishing piece of food.

The western world connection to nature is an experience where you can look but not touch nature. Sobel [78] recounts several reasons why we have lost our real connections to nature:

> "...—urbanization, the changing social structure of families, ticks and mosquito-borne illnesses, the fear of stranger danger. And perhaps even environmental education is one of the causes of children's alienation from nature."... "Much of environmental education today has taken on a museum mentality, where nature is a composed exhibit on the other side of the glass. Children can look at it and study it, but they can't do anything with it. The message is: Nature is fragile. Look, but don't touch. Ironically, this 'take only photographs, leave only footprints' mindset crops up in the policies and programs of many organizations trying to preserve the natural world and cultivate children's relationships to it."

Western world connection to wildlife is mostly seen through a zoo lens. Zoos are designed to isolate humans from endangered and dangerous species [79] but still give humans an entertaining experience. The first "zoo" was built more than 4,000 years ago in Egypt but does not fit today's definition of a zoo. It was a private collection of animals kept in a garden by Queen Hatshepsut. It was not available for the general public to visit. The term zoo was first coined by the London Zoological Gardens when it opened its doors in 1847 to a paying public because it needed help to defray its costs (www.zsl.org).

Most historians recognize the oldest built zoo as the one found in the Schönbrunn Palace in Vienna, Austria. It was built in 1752 by the husband of Maria Theresa of the Habsburg monarchy as a summer residence for the royal family. It was not originally built for the general public to visit but as a way to impress other aristocrats. Now it is visited by tourists who vacation at the Schönbrunn Palace. When the zoo was first built at Schönbrunn Palace, the www.swlearning.com/pdfs/chapter/0538439513_1.PDF1.

> "Empress did not like having dangerous, noisy, and smelly wild animals of prey close to the Summer Palace, so there were no bears or tigers in this first zoo. Under Joseph Ⅱ, the director of the zoo was sent to Madagascar and Southern Africa in search of exotic animals. Since 1772, over 35 elephants have called this zoo their home each year!"... "Today, there are 3,000 animals... at Schönbrunn.... The zoo... is one of the World cultural Heritage Sites of the United Nations Educational, Scientific, and Cultural Organization (UNESCO)."

Two of the authors visited this zoo at Schönbrunn Palace some 25 years ago. It was not a pleasant experience. The houses or cages were small and looked quite cramped. The surrounding area around the cages did not have any green vegetation to improve the aesthetics of the habitats.

At Schönbrunn Palace, the animal habitats were not suitable for animals used to living in tropical climates. Monkeys from the tropics lived in concrete cages in a cold Vienna climate with little protection from the elements. Monkeys were confined to small spaces that were inadequate for their needs compared to what modern zoos offer today. Monkeys sitting in their cages did not look healthy and appeared to have visible skin problems. They paced back and forth in their cages similar to what is seen for wild animals confined to small spaces. Even tropical parrots were displayed in cages in the open.

The zoo at Schönbrunn Palace was built for its architectural elements and not to satisfy the needs of animals housed in its facilities. Today, this zoo is a UNESCO site because of its historical and architectural roots but not because of its animal habitats.

There is a historical context for why the Schönbrunn Palace zoo was built the way it was. This zoo was built in the late 1700s when there was little understanding of how zoos could be designed to provide a suitable habitat for animals that lived there. Early built zoos were small and not built to mimic the natural habitats of the wildlife.

Zoos began to be built in most small towns in America after World War Ⅱ. Western societies already had good experiences in building botanical gardens some 300 years before zoos were built. These were places that societies could reconnect with plants and animals. All these encounters, however, were very well controlled and reflected human views of nature. The mundane or common plants and animals were never part of these exhibits. Botanical gardens were more interested in growing plants exotic to the region and not informing communities on plants growing commonly in their region.

Every small town scattered across the U.S. used to have a zoo even if it was small. Clovis, a small town in northeast New Mexico, had a zoo even though it only housed a few animals. If you were viable small town and wanted to attract people to settle down in your town because it was a good family town, you had to have a zoo. Of course, animals shown in these zoos were collected from exotic places since no one is going to want to go to a zoo to see a cow or a domestic cat or dog. Zoo animals had to be unique so that visitors would continue to make frequent trips to the zoo with their children. Elephants, large cats and giraffes were especially interesting to look at. The popularity of circuses coming to small towns also attests to human fascination in seeing exotic animals performing for their benefit.

Today, zoos continue to be important entertainment opportunities for people living in cities. Its role in society is changing and appears to provide a way for children of today to experience a small part of nature. The western world is concerned with nature deficit.

Nature can't seem to compete against technology which is easier to play with, perhaps more attractive and cleaner. If children played in nature they would experience more messy connections to nature since it rains or snows in nature. Nature can be muddy or very wet. One would potentially become quite muddy after jumping into a pond for a swim.

We do not have sanitizers located next to ponds that would spray clean and remove germs from children as they emerge from the water. It is much easier to go to a zoo. You do not need to clean up after a zoo visit. Zoos have become locations for society to view a sanitized nature where animals entertain us. Your safety is never a question that comes to your mind when you visit a zoo.

A zoo was designed to provide its visitors sanitized views of nature. It kept the visitor at a distance from the reality of animal wastes, dirt or grime that is found in nature. Since fewer Euro-American descendants of the earlier colonialists hunt today, their experience of nature is this sanitized and tame view that zoo's can offer.

There is no danger in going to a zoo because the animals are behind cages and are unable to touch you. Zoos would be a decidedly different experience if animals were released into the zoo grounds so humans could play out a game of hide-and-seek or a survival game with these animals, e.g., either the animal won when you became their meal for the day or you killed the animal and won yourself a trophy to hang on your wall. This is not going to happen. Humans do not want this kind of experience even if this story has been written in science fiction books.

Many of the animals that are popular exhibits in zoos are natural predators on humans in the wild and would attack humans if given half a chance. So the predatory behavior of zoo animals is not allowed to become part of the zoo experience because of human safety concerns.

### 5.2.3 Today's Nature: Bounded Larger Artificial Landscapes

In the early 1900s, zoos began to dramatically change. People became more familiar with animals from all over the world. Whether this was from the books they read, the nature programs they watched on TV or actual travel around the world is not completely certain. Whatever caused it, people became bored when they went to their local zoo and saw an animal simply lying in their small cage without much sign of activity. Others were greatly disturbed by the sight of these caged animals that they knew should be running or climbing in a more natural setting. Ideas began to circulate about how the animals should be taken care of and displayed in settings that were closer to how they would live in the wild.

In 1907, the German entrepreneur Carl Hagenbeck founded the Tierpark Hagenbeck in Stellingen, now part of Hamburg. It is known for being the first zoo that used open enclosures where animals were surrounded by moats, rather than the barred cages used in most other zoos. This improved the lives of the animals and came closer to approximating their natural environment. This also enhanced the education of people viewing the animals since they saw them in something closer to their "wild state". In 1931, the Zoological Society of London opened the Whipsnade zoo in Bedfordshire, England. This zoo was the first wild Animal Park, or open range zoo, and contains 243 ha of land for its animals. This was a still very controlled experience even though it appears that wildlife and humans have the ability to touch one another since barriers or the borders are hidden.

Relative new terms for zoos have been coined for the 20th century. They are now called "conservation parks", "bioparks", open-range zoos, or safari zoos. Adapting a new name is a strategy being used by zoo professionals to distance them from the stereotypical, cruel and nowadays highly criticized zoo concepts that existed prior to the 20th century. These parks and zoos cover very large areas that allow the animals to run or climb and animals that lived in herds could be together as they would in the wild. For the most part, this type of zoo has fewer animals. Animals are kept in larger outdoor enclosures using moats and fences to contain them from humans rather than cages. Most Safari Parks allow visitors to drive through them and come in closer contact with the animals. Sometimes visitors get closer than they might want.

When ecology became a matter of more importance to the general public in the mid 1970s, a number of zoos started to consider making conservation their central role. This conservation now includes special breeding programs for endangered animals. The position of most modern zoos in North America, Europe and Australasia is that they have wild animals on display for the conservation of endangered species. Research and education purposes or the entertainment of visitors is far from their primary function.

Starting in the 1970s, the environmental movement contributed to the dialogue on what role zoos should have in our society. Some question the value of experiencing or seeing animals under the artificial conditions found in zoos. Zoos are a decidedly western European experience and one that most indigenous communities would not establish. Native Americans have not separated wildlife or nature into discrete and sanitized experiences. Partly this is due to the fact that they are busy surviving and hunting wildlife for food. Wildlife has spirits and should be revered and not be confined into a bordered experience.

Today, the role of zoos is starting to change because of the many challenges zoos face because of their small size. Zoos will never be able to sustain large viable population of individual species because they are too small [79]. Zoos find it difficult to breed animals without genetic defects or keeping disease from spreading and killing animals confined to smaller spaces.

Kaufman [79] recounts how the Copenhagen Zoo had to kill two leopard cubs that were about two years old "whose genes were already overrepresented in the collective zoo population". This zoo "annually puts to death some 20 to 30 healthy exotic animals—gazelles, hippopotamuses, and on rare occasions even chimps" [79]. The rationale given for this practice was that most of the offspring would "die from predation, starvation or injury" [79]. Zoos have been criticized for euthanizing children of animals that they breed. However, zoos do not have funds to maintain large populations of many species on its grounds. Zoos defend their actions as the need to allow animals to raise their offspring so they can experience parenting, even if they ultimately euthanize them [79].

## 5.3 Native American Model: Borderless Nature

*I do not think the measure of a civilization is how tall its buildings of concrete are,*

*But rather how well its people have learned to relate to their environment and fellow man.*

~ Sun Bear of the Chippewa Tribe ~

The quote by Sun Bear articulates well the Native American view of nature. Nature is not compared to the human "built" environment, e.g., our castles, churches, fortresses, skyscrapers, etc. For Native Americans, how civilized you are depends on how well you relate to nature.

Native Americans are connected to nature in a multitude of ways and not how you can individually optimize the production of products with revenues generation potential from nature. The story written by Rodney Cawston in Box 16 shows the multitude of ways that an Indian growing up learns to connect to nature. His story is from the Pacific Northwest U.S.

**Box 16: Rodney Growing Up with Nature**

With every tribe, gathering and hunting knowledge is tied to cultural beliefs, values, and behaviors. Native people respect every rock, hill, plant and animal as having a soul and a spirit. Tribes have cultural rules or laws that are exercised when gathering or hunting. Historically, Native people always believed that everything extracted needed to be used and not wasted. Large game animal parts were incorporated into an array of cultural objects: tipis, adornments, baskets, ceremonial costumes, clothing, cosmetics, dyes, foods, games, household utensils, medicines, musical instruments, tools, toys, and weapons. Mountain sheep horns were fashioned into spoons. Nimal sinew and ligaments were boiled down to make glue. My grandmother told us that in her early days they would cook much of what we now throw away. The intestines were stuffed with meat and roasted; the windpipe was finely chopped into a meal and added to soup.

Native peoples in the Pacific Northwest had a variety of large game animals as resources, including deer, elk, moose, mountain goats, mountain sheep, and bears. Deer was one of the most important resources for many tribes. Historically, Native people knew the behavior of both white tail and mule deer through living within their natural world. They shared watering places, recognized identifiable signs such as tracks, marks on trees, and crushed vegetation where the deer would bed down. I used to go out hunting with my father and great uncle and they could cleverly whistle like a deer to learn of their locations. Native people also used fire, traps and snares to capture herds of deer. Native people also are keenly aware of cyclical changes of large game animals, especially deer. It was very important that they knew their seasonal migration patterns and reproductive cycles. A doe carrying or raising a fawn was never taken. My father could look at a doe and knew whether she was pregnant or had a fawn. He could also differentiate a doe from a buck when the buck has no horns. When a deer was killed my father would hide or bury the parts of the deer that was left

(Continued)

behind. He would always say that this was a form of respect so that the deer that would go back looking for their lost one wouldn't feel bad. Out of this respect, every part of the animal was used.

Before going out hunting many tribes will take a sweat bath and pray. It is important to Native people that you have a good heart and good mind when hunting or gathering. It is believed that if you treat the deer with respect they will show themselves to you. Many tribes have first food feasts where they will pray for their traditional foods as an expression of giving thanks for these resources and that they remain plentiful. When a young man kills his first deer many tribes have ceremonies or cultural practices to celebrate this event. It is the belief of Native people that we coexist with the world around us. There are stories, songs, and handed down knowledge that teach us to act respectfully when extracting these natural resources. We are taught not to ever kill anything needlessly and to never kill anything on Sunday. The seventh day with many tribes is sacred and one of our most important teachings is that everything should be able to perpetuate itself for one day.

(Confederated Tribes of the Colville)

### 5.3.1 American Tribes: Nature, Sense of Property Is Culture-based as Told by Mike[1]

> *You must teach your children that the ground beneath their feet is the ashes of your grandfathers. So that they will respect the land, tell your children that the earth is rich with the lives of our kin. Teach your children what we have taught our children, that the earth is our mother. Whatever befalls the earth befalls the sons of the earth. If men spit upon the ground, they spit upon themselves.*
>
> ~ Unknown Native American quote ~

When I entered into First Year Law at the University of Washington in 1972, the classes would state that Real Property law is like a "bundle of sticks", it is actually a collection of rights for lands. Property has spatial aspects to it. There is length and width, the square footage, acreage, square miles, for example. There is also up and down aspect, the mineral and subsurface water rights below

1 Confederated Tribes of the Colville.

and also the air rights, which go vertically up, which may have visual issues or even air quality factors or aircraft rights which can interfere. Property also has a time aspect. You can rent a hotel room for one night. Or you can rent a room or pasture or office for a month. Or you can enter into a lease. Or you can outright purchase lands. If you want "total" ownership, this is fee simple land. You own the physical aspects and you own the time aspect, there is no limit on time, it is yours forever. But as property class progresses, you will be taught that there are no absolutes in law, there are always exceptions, shades of gray and the devil is in the details. Thousands, if not millions of land transactions take place every day for rents, leases, and purchases in the U.S. There is also a role for the state too, but introductory law does not usually dwell on it too much. If a freeway needs to be built, for example, the state has authority to condemn land for the good of the whole. But most of the emphases are on the rights of the individual. Individuals buy land for business or for homes. Corporations can buy land too of course. Transactions get filed with the local county governments. Tribes operate much differently.

If the subject of Native American land ownership ever comes up at all to the lay person, it usually hits on a couple of aspects. One day you will often hear that the land was not actually being used anyway or that the land was vacant and there for the taking. Or maybe the trade of $24 of beads and trinkets for Manhattan Island in the Colonial days will be mentioned, and then there will be a couple chuckles as to the ignorance of the Natives and the discussion moves on to something else. Lands in the Pacific Northwest were rich with natural resources on the land and on the water. Salmon were abundant. The land supported edible plants such as camas roots, bitteroots, wild carrots, asparagus, and many others. There was an abundance of wild game such as elk, deer, and moose. Often, these foods were ready to gather only at certain times of the year. Over time, the tribes made the seasonal rounds and knew when to gather and harvest nature's foods. Their concepts of land ownership were influenced by the food gathering patterns, by when they were ready to harvest. Native Americans would come in and harvest and then leave to the next site. Their culture evolved to be highly mobile in order to take advantage of the natural abundance. It made no sense to make permanent settlements at any one site. The bounty of the land was also shared, there was a communal aspect. Roots, berries, deer, elk, they provided for the family, but it also normal for a share of the foods to be distributed throughout the community, the tribe. The poor, the elders, and the sick were fed and taken care of by the tribe. The earth was viewed as the mother. You respect your mother. You do not abuse your mother. When the U.S. government came in during the late 1800's to convert the Indians into agricultural people, the Chiefs would say, do you expect us to plow up the ground? That would be like dragging a knife across our mother's heart. You also do not sell your mother.

The Columbia River begins in what is now British Columbia, Canada and flows southwest into the Pacific Ocean crossing the States of Washington, Idaho, and Oregon. The River was the dominant feature in the landscape and life of the Columbia River tribes. It provided salmon in great abundance and the people built their economies and cultures around the great River and its food supply. But the river also carved its way through mountains and these mountains and valleys provided a rich variety of terrain and vegetation and wildlife that all provided a living to the thousands of Native Americans, whose life styles had been stable for at least 10,000 years since the last ice age.

As a young child, I was raised by grandparents, William and Mary Marchand. They were Arrow Lakes people and members of the Colville Tribes in Washington State. Grandfather was an entrepreneur. He had a sawmill, raised hay and cattle, and did other businesses. Grandmother was very traditional, she liked to live the old ways, gathered roots and berries, dried salmon, and preferred living in tents during the warmer months and lived very much like our people had for thousands of years. We spent many nights around the campfire and she would tell stories and teach us. She was the daughter of the last hereditary Chief Aurapahkin, and she knew a lot of stories. In winter, we lived in permanent homes, like most Americans. My grandparents actually had three homes and owned a lot of property. She followed the roots and berries and salmon around throughout the seasons. In spring we camped at Owhi Lake to catch the trout. When the roots were ready, she camped in the root fields. When the salmon runs came, she camped at the mouth of the Okanogan River where it meets the Columbia River. Prior to 1942, she camped at the once great fishery at Kettle Falls, but it was inundated by the Grand Coulee Dam. In the fall, she camped in the mountains near Sherman Pass to pick huckleberries.

Every year she would follow the roots, the berries, the trout, and especially the salmon. Distances covered would be hundreds of kilometers in the course of the year. They would load up a couple of cars with camping gear and provisions and head for the outdoors. By her lifetime of experience, she knew when it was time to go camping and she would return to the same places each year. It was an annual routine that she followed all her life. The food was gathered at the camps, and then it was either dried or canned usually. Stored for the winter months when it was eaten. There would be big feasts for the clan in the spring, summer, and fall seasons. The men would sometimes go off by themselves and hunt too. Then in spring the cycle would repeat. This is how the Columbia River Indians lived for thousands of years. Her people were from the headwaters of the Columbia River, the Arrow Lakes people. But all the downriver tribes had similar patterns of life. When she was younger they traveled the same routes on horseback.

Grandmother's view of the land and ownership were different than the European view. She used the land only at certain times of the year, such as when the fish were ready to harvest or when the berries were ready to pick. Some of the lands were owned by the tribe, some owned by the Federal or State governments, but she considered them her traditional gathering grounds. Over thousands of years, regional boundaries between tribes had been established, sometimes in times past, there were wars and skirmishes between tribes, but over time these boundaries were established and respected as tribal territories. At any given moment in time, however, from the outsider viewpoint, it was likely that any particular lands appeared vacant.

The first non-Indian visitor to my grandmother's people was explorer David Thompson. His expedition came through the Columbia River in 1803. He met with my grandmother's ancestor Chief Gregory at Kettle Falls and witnessed the Arrow Lakes people fishing. He noted they were pulling 2,000 salmon out of a single fish trap in one day. The people placed great baskets beneath the Kettle Falls and the traps filled up with salmon who were trying to jump up the falls. This practice continued right up until the 1930's. My grandfather, William Marchand was arrested by the State of Washington for using these same basket traps, not approved by the State. The fish were shared with the tribe. The poor and elderly and the sick were taken care of. The surplus was traded. It was big business.

At one time there were bales of dried salmon stacked all around the Kettle Falls. The tribe was wealthy. Thompson then went downriver and passed by the mouth of the Okanogan River and reported fishing camps and horse racing. Thompson said the long houses they lived in were 100 meters in length. As a young boy in the 1960's, this same mouth of the Okanogan River fishing site is where my grandmother would put her tent each summer. This ancient fishing site at the Okanogan mouth was also later inundated by another dam in 1964, named Wells Dam. So that is the history, Native Americans getting continually displaced by U.S. developments and interests.

### 5.3.2 No Walls: Active Landscape Management, Nature Not Wasted

*If you continue to contaminate your own home, you will eventually suffocate in your own waste.*

∼ Lakota Proverb ∼

Prior to the arrival of the European colonialists, tribes had customary land areas where they would harvest resources. They did not establish formal boundaries defined by walls to demarcate lands where they would harvest their resources. By contrast, Europeans are superb wall builders and broke the landscape up into discrete units for very specific purposes. When the Romans could not control the marauding barbarians of northern Europeans, they built the Hadrian's Wall to keep these people from attacking Roman settlements. Building walls has served western Europeans well and allowed them to control nature in tidy and discrete blocks of land use. The people who lived in these blocks liked nature with borders. No structures built by Native Americans can be found today that split the lands up.

Even though Native Americans were regarded as "hunter/gatherers", they had extensive husbandry practices to encourage growth of valuable resources [71]. How do we know the tribes were active forest resource managers? Is it possible that the evidence of fire was just accidental lighting strikes in forests during lightning storms? Or perhaps it was just natural occurrences of lightning strikes? Evidence supporting active management of lands and wildlife is derived from [71]:

- Early accounts by missionaries and fur trappers who described these activities.
- Natural fires generally occur in late summer and early autumn when plant moisture is reduced and dry undergrowth has accumulated. By contrast, human-induced fires are set primarily in the spring when it is possible to stimulate maximum plant growth, and at times that the fire could be easily managed.

It is commonly accepted today that Native Americans managed the land and the animals that lived on these lands [71]. Reports suggest that Native Americans managed the ungulate populations. This is supported by the observation that it was a rare sight for American settlers to see a moose after 1825 or an elk in Yellowstone National Park after 1835–1872. Today this is a very different situation with Yellowstone National Park having at least 100,000 elk living inside the park borders. The beaver populations were also quite high during the early years when Euro-Americans first arrived in North America. The high beaver populations also support the management of elk since the preferred wood used by beavers is young aspen saplings that are also frequently grazed by elk [71]. Bonnicksen et al. [71] wrote

> "American Indians protected habitats and promoted biodiversity in plant and animal communities by keeping ungulate numbers low... Prior to the early 1800s, for example, millions of beaver (*Castor canadensis*) occupied lush riparian zones throughout the West. Beaver were so abundant that in 1825, Peter Skene Ogden's party trapped 511 beavers in only five days on Utah's Ogden River. In 1829, Ogden also reported that his fur brigade took 1,800 beaver in a month on Nevada's Humboldt River ... Yellowstone too once contained large numbers of beaver, but that species is now extinct on the park's northern range.

> Without American Indian hunters, the park's burgeoning elk population has nearly destroyed the willow and aspen communities that beaver need for food and dam-building materials ... So American Indian hunting benefited all species by preventing habitat destruction by large populations of ungulates."

This of course all changed when the Euro-American colonialists arrived. They began to demarcate land into areas with borders. They began to force Native Americans onto reservations and to remove them as the landscape managers. Treaties set boundaries where there had not been boundaries beforehand. Today these forced settlement of several tribes on one reservation resulted in the merger of tribes that historically did not live together. This has created artificial boundaries for tribes that they are still attempting to deal with. Today, tribes have to address who has rights to the wealth of the resources generated on the reservation. Who is a "real" Indian? Who belongs on the reduced land base and who should benefit from whatever resources are harvested from these lands?

Even though some of the Native America tribes were superb agriculturalists, western European colonialists wanted them to practice "real" farming as defined by Europeans. European colonialists justified taking Indian lands since the American Indians were not using the lands at the intensity that made you "civilized", i.e., the lands appeared to be vacant and not being managed. Of course, this was not the case but Europeans did not recognize a land-use that differed from theirs. The European model of land use is to transform or convert nature into something not resembling nature. Nature needed to be controlled.

There were fundamental differences in how agriculture was practiced by the Euro-American settlers and the Native Americans. Native Americans farming practices followed traditional ecological knowledge where nature provides food without needing to be converted into industrial scale farming fields. Nature did not need to be totally transformed to feed people. Europeans, however, needed more intensive agriculture to produce enough food to feed the growing populations in a rapidly industrializing Europe. As recounted by Maier et al. [12], Europe could not have industrialized if they had been dependent upon growing the food crops that already grew on the European continent. European industrialization was possible because of the more than 300 food and pleasure crops that were taken back to Europe by the early colonialists [12]. These food crops grew in Europe and had higher yields compared to what had historically been planted in Europe.

The quote at the beginning of this section reflects an aspect of Native American views of nature. It shows the importance of not wasting nature or its resources. As the Lakota proverb mentions, you will suffocate in your own waste if you convert too much of natures capital into wastes. Native Americans are not wasteful and use every part of any resource they collect. The western world is still trying to figure out how to deal with the volumes of wastes that industrialization produces.

To be sustainable we need to know when our consumption of resources is wasteful or when your continued use will further aggravate its scarcity. Native Americans knew when scarcity was possible and adjusted their behavior accordingly instead of waiting for the resource to become scarce before working on solutions. This is similar to the quote about knowing when the well will contain less water and not to know this fact when the well is already dry. Then it is too late. When the well is dry, it is too late to resolve this problem. You have also lost all your options if you wait too long to work on a solution. It is difficult for western societies to proactively work on solutions. It is easier to work on a problem when it becomes a problem. This is partially the result of too many problems that need to be resolved and the difficulty of prioritizing which problems to tackle first.

**Coyote Essentials**

- A person's culture determines their view of what constitutes a home and what may be considered personal property. Many tribes consider wherever they are currently located as home while Europeans tend to consider their home as a specific location or place.
- Native Americans believe that the Creator made the earth as a place for them to live and thrive. They do not view the earth as a resource to be used as a commodity.
- Empty/Vacant undisturbed land has significant ecological, spiritual and cultural value.
- Tribes are willing to adopt most new technologies as long as they do not negatively impact or conflict with their cultural values.
- Europeans do not consider/take into consideration cultural values in determining which technologies are acceptable. Europeans do rely on technology to resolve societal and economic problems.
- Technology is compatible with nature if resources are used efficiently and wisely, and managed lands are not "decoupled" from nature/retain their overall natural components or are restored to natural condition.

# Chapter 6

# Portfolio Element II: How to Make Practical and Realistic Decisions

*Never trouble anyone regarding his religion—respect him in his beliefs, and demand that he respect yours.*

∼ Tecumseh (1768–1813) Shawnee Indian chief—led intertribal alliance resisting white rule ∼

## 6.1 To Become Sustainable Don't "Throw Out the Baby with the Bathwater"

Sustainability requires a long-term view. Unfortunately most western world people view problems as something that needs to be immediately resolved and not something that might exist for decades. People become tired of a problem if it cannot be resolved quickly. There are too many other pressing issues that need attention to continue to work on a tricky and unsolvable problem.

We contend that this short-term view is driven by tools that allow rapid assessments of any problem. Resource economics tools, and the assumption that technology will ultimately solve any problem, appear to solve any problem. There is a false sense of security that technology will always produce a solution. Technology has always come through in the past so why would it not do the same in the future? To be sustainable, however, technology should only be part of the solution. A purely technological solution will make societies and environments less resilient, not more resilient.

We contend that western European tools and decision-process models are applied inappropriately at many locations. This happens even when the solutions are "foreign" to the local context and therefore difficult to "naturalize" for the local situation. We do not to say that western European tools and decision-processes have no role in today's world. But they are being implemented without including the local knowledge that would make them place-based and better to be able to provide a solution that works locally. If decisions are made by foreigners who can gain economically from the decisions, it is easy to justify using a business model supported by knowledge developed in geographically distant locations. The context of the local and foreign place appears to be similar so it is easy to justify using this "foreign" knowledge of a local place. Such an approach ignores the people who live on the land and have memories of how this land works. These people Know these decisions will impact their local survival.

A business model approach to decision-making shifts the focus to capitalize on the efficiency of technology to maximize profits and economic returns (see section 7). It appears to be a win-win situation. Decisions can be made quickly when supported by economic analyses that suggest less costs and more revenues will be generated by a decision. Because of this, it is difficult to slow down the decision-process and to include non-economic decision models in the assessment process.

When economic criteria drive the decision process, building large-scale centralized bio-energy facilities is always more economical. Economists never consider decentralized facilities to be economical to build in rural areas. This happens even when biomass supplies become prohibitive to transport to large centralized facilities. Under normal conditions, centralized facilities consuming biomass are typically most economical if the resource supplies are within a 100 km circle around a wood energy facility. Beyond this distance, it is not economical to transport a bulky and high density material to a centralized facility. Distributed energy systems are economical in rural and remote areas because of the efficiency of scale [80]. These choices are not being made because economics is the primary tool being used to evaluate whether a business should be large or small scale. The large scale always wins despite the distribution supply problem.

Interestingly, a large centralized facility is less resilient and economically viable after a natural disturbance. Businesses should factor in "boom and bust" cycles since they are less viable during an economic downturn. For example, following a hurricane, the transportation system is broken. It will be more difficult to keep large centralized facilities operational during these times. So the economic analyses need to be made during the business boom as well as its bust cycles. A boom and bust cycle buffers social and environmental factors in any business enterprise. Taleb [81] wrote

> "Projects of $100 million seem rational, but they tend to have much higher percentage overruns than projects of, say, $10 million. Great size in itself, when it exceeds a certain threshold, produces fragility and can eradicate all the gains from economies of scale. To see how large things can be fragile, consider the difference between an elephant and a mouse: The former breaks a leg at the slightest fall, while the latter

is unharmed by a drop several multiples of its height. This explains why we have so many more mice than elephants."

Despite the recognition that social, economic and environmental issues have to be examined in all decision processes, it has been problematic to expand economic analyses to formally, not anecdotally, include other factors. Environmental and social factors are complex, last decades and add competing solutions to a decision process [82]. This would definitely slow down the decision process since the information in these areas is less concrete. It is easier to separately analyze the social and environmental factors from the economic assessment. This addresses this problem as a cosmetic solution [1]. Since many funding organizations expect to see economic data, it is the most common tool used to explore the viability of different business options. This situation will be difficult to change because today's business leaders drive economic growth and determine what tools are used.

We feel that it is important to identify when a problem can be assessed using economic tools and when an economic tool is used to solve the wrong problem. It is worthwhile to go back to the definitions of the terms ecosystem and adaptive management made by John Gordon [74]. These terms help to identify which tools might work best under what conditions. Economic tools work best when decisions are being made on land scarcity issues. The choices are clear since the production capacities of the land to produce any products are finite. Economic tools do not work well when there is a need to factor in uncertainty or scarcity of scientific knowledge and information. Under this situation, the wrong information can be easily used to calculate the costs and benefits. Most resource problems decisions are made under the latter umbrella and not the former.

When scarcity of knowledge is not acknowledged, business solutions may exclude relevant information that may direct you to pursue a solution that is not economically viable in the long-term. This approach can cause you to "throw out the baby with the bathwater" or as it was originally spoken "das Kind mit dem Bad ausschütten" [83]. Murner used this saying as a proverb in his satirical book that he wrote in 1512 and depicted on a wood cutting shown in Figure 6.1 [84]. It refers to German villagers at the beginning of the 16th century taking turns to bath in

Das kindt mit dem bad vßschittẽ.
Ein narr der meint es sy nit schad
Das kindt vß schitten mit dem bad
Vnd sy so gůt in die hell gesprungen
Als mit tüttschen dryn gerungen

Das ist in aller welt gemein
Das kein vnfal kumpt allein
Er bringt mit im vnglücks genůg
Das mancher narr nie wardt so clůg
Wie er sich sol vß vnfal ringen
Wa man in wolt von sym gůtt bringen
Vnd felt im zů ein widermůt
Den im ein narr vff erden thůt
So henckt er sich dañ selbs darzů
Vnd schlecht das kalb vß mit der ků
Vnd schit das kindt vß mit dem bad
Zů vnfal macht im selber schad

**Fig. 6.1** Wood cutting made by Murner for his satirical book, 1512 [84].

the same tub of water. Men had first rights to bathe followed by women, children and then finally babies. The idea was that by the time babies were being bathed, the water would be so grey and dirty that there was a chance that a baby could accidentally be thrown out with the water. This action is not resilient because it causes a bad thing—dirty water—to be thrown out with the good thing—baby.

Today it is recognized that resilience is the ability of the social and natural systems to adapt to change and to continue to maintain all the components of an ecosystem and its connections and functions. If there is a scarcity of knowledge, there is a strong possibility that wrong variable is used in the decision process. In cases like this, it is easier to accept knowledge provided by special interest groups since they are able to make a good case for their values without revealing the selective nature of this data. They appear to espouse the truth. Today, it has become more difficult to include the special interest groups, who have a vested interest in the outcome of any decision process, but not give up control of the process to them.

## 6.2 Leave Your Individual Biases Outside the Door

### 6.2.1 "False" Indian Stories

Since early American Indians were hunter/gatherers, the perception is that people who live by collecting from nature will spend most of their time on searching for food. People assume that hunter/gatherers have no time to create cultural artifacts since survival is of the utmost importance. Survival meant foraging for food. It did not mean crafting baskets or decorative ornaments that had no utilitarian uses. These misconceptions still exist today. The reality is that for hundreds of years, Native American hunter/gatherers spent less time on acquiring the essentials for survival than Europeans living at the same timeframe. Miller [8] writes that most Native Americans did not spend all their time on collecting food and pursued many different cultural activities for more than half of their awake hours. Miller [8] writes

> "For many Indians, their work and ingenuity provided an ample source of life's necessities and their 'economic year'—that is, the time it took them to produce all their annual food and subsistence needs—was only about four to five months. That short economic year left plenty of time for leisure, culture, and ceremony. In contrast, most Americans today

> have an economic year of 50 weeks because we only get two weeks of vacation."

Physical anthropologists studies on the origins of agriculture support the idea that hunter/gatherers did not have to spend all their time on searching for food to remain healthy. By contrast, they write that as agriculture developed people became smaller and were prone to more diseases than their hunter/gatherer forbearers [85]. By contrast, anthropologists suggest that once people became full-time farmers, they spent most of their time on working at surviving. They were not crafting decorative artifacts that today's generation can buy as antiques. They did not spend time on becoming creative; they survived by plowing fields and feeding their animals. This was hard work that started before the sun was visible in the horizon and continued until the sun set.

It is ironic that the western Europeans wanted to civilize American Indians and to force them to practice the European type of agriculture (see section 2.1.1). The European agricultural practices did not leave time for people to become creative and to use the right side of the brain (see section 3.3.2). By contrast, the hunter/gatherers had more time to spend on artistic activities and led a healthier life. American Indians produced abundant examples of arts and crafts, e.g., blanket weaving, bead work, pottery, baskets and so on. Today Americans spend thousands of dollars for buying these Indian artifacts when they show up in Antique Road shows or at auctions.

This continuing fascination with anything Indian is apparent from the high price paid by a private collector for Chief Joseph's war shirt. This shirt was sold in an auction on July 21, 2012 and its sale is recounted below by Martin Griffith for the Associated Press:

> "Chief Joseph wore the shirt in 1877 in the earliest known photo of him, and again while posing for a portrait by Cyrenius Hall in 1878. That painting, which was used for a U.S. postage stamp, hangs in the Smithsonian."
>
> "The poncho-style war shirt was made of two soft skins, likely deerskin. It features beadwork with bold geometric designs and bright colors. Warriors kept such prestigious garments clean in a saddle bag on their horse or carefully stored while in camp, to be worn only on special occasions, ..."
>
> "The photo and portrait showing the war shirt were made shortly after Chief Joseph led 750 Nez Perce tribal members on an epic 1,700-mile journey from Oregon to Montana in an unsuccessful bid to reach Canada and avoid being confined to a reservation. They were forced to surrender in 1877 after U.S. troops stopped them about 40 miles south of the Canadian border."

Today, perceptions of non-Indians on what it means to be an "Indian" are fabrications and myths developed by Hollywood, other media and movie outlets. In Indonesia, movies have depicted Native American tribes as loving war so the average person in Indonesia thinks that American Indians like to fight all the time. This is not correct but movies keep this idea alive and at the forefront of the movie going Indonesian public. This is a romanticized view developed by the western world because it sells movies. If American tribes were depicted in movies constructing crafts like baskets, no one would pay money to see such a movie.

American Indians seemed to be frequently confused with people coming from India. This early confusion was probably warranted since the early explorers really did not know where they had landed. It is said that Columbus to his dying days never realized that he had landed in a different world, the Americas. These old confused ideas became the "truth" about American Indians and have been propagated to this day. Wilton [83] (2009) describes the origins of the word Indian as being "racially motivated, false etymological origin". The media propagated view suggests American Indians are either "savages" or "noble savages". It does not repeat what is written in etymological dictionaries that attribute American Indians with "honesty or intelligence". If one looks up the etymological dictionary for Indian [86], it is revealing to see that it has a very different meaning from how it was originally used. It has been corrupted today by comedians like Carlyle who called Indians the "Red Indians":

> "inhabit of India or South Asia," c.1300 (noun and adjective); "applied to the native inhabitants of the Americas from at least 1553, on the mistaken notion that America was the eastern end of Asia. Red Indian, to distinguish them from inhabitants of India, is first attested 1831 (Carlyle) but was not commonly used in N. America. More than 500 modern phrases include Indian, most of them U.S. and most impugning honesty or intelligence."

Native Americans have been characterized as either bad or good depending on who is having the conversation. They are not described as having the two aspects of the coyote that is a blend of the good and bad depending on the situation. Probably every person in this world is more coyote like but no one wants to admit that they have both good and bad thoughts and behavior attributes. The goal is to be good. No one will admit to bad behavior. In the western world, a person always succumbs to the good voice or the angel who is talking to you on your right shoulder and telling you what to do. No one would admit that he/she listened to the bad voice or the devil talking to you on your left shoulder.

Unfortunately once society builds a view or an idea about someone or a group of people, it is very difficult to change this view. This is well summarized in a New York Times opinion article written by David Treuer—an Ojibwe Indian [87]. He wrote

> "Growing up as I did, on the Ojibwe Leech Lake Reservation in northern Minnesota, it was patently obvious to me that Indians came in all different shapes and colors. I'm fairly light-skinned and have been told many times that... I can't be an Indian, not a real one."

David Treurer mentioned the old American narrative on who is good or bad when he wrote [87]

> "where Indians are noble and dark and on horseback, and just as divorced from the textured complexity of the American experience; one where the good guys are broad-chested and the villains twirl their mustaches..."

This inaccurate view of Indians and what they should look like is part of the past narrative that continues today. What is interesting about this is that if you do not fit our pre-conceived notions of what you are supposed to look and act like, our response is to say that you are not a "real" Indian.

This idealized image of the Wild West and the role Indians play in it have continued since the first European colonialists landed on North America. Seymour [87] summarized how in the late 1880s English gentlemen tried to build a traditional hunting club in the prairies of present day U.S. They did not really want the "*real*" Wild West but a facsimile of what they had back in England. Seymour wrote [88]:

> "Back in the late 1880s, Runnymede was the spot where a group of English colonists decided to live out the fantasies they had read about in the exotic cowboy tales of Wild Bill Hickok and Calamity Jane. Known derisively as 'remittance men' because of the comfortable allowances they regularly received from their daddies back home, these young toffs had been lured to Kansas by Francis Turnley, a wealthy Irish landowner's son from County Antrim who had gambled his future on the chance that a new railway line would lay its tracks straight through his personal fief: Runnymede.
>
> Turnley worked hard to recreate upon the prairie a perfectly pukka England. Runnymede offered not only polo matches, tennis tournaments and cricket, but a splendid opportunity for aspiring cowboys to don their buckskins and fire off six-shooters as they swaggered down Main Street. The new arrivals from England loved Runnymede. Estab-

lished locals, especially the pious German-American settlers at nearby Harper, didn't think much of the antics of these grandee colonists. ..."

In the 1880s, English gentlemen romanticized what western cowboy life was all about. This life had no reality to it but it was real to the gentlemen play acting the "cowboy life". This life did not include Indians or the hardship of living in the Wild West. This was a decidedly unrealistic view of how difficult it was to survive in the western U.S. Since the new federal government needed to attract settlers out west to retain their control of these lands, who were they going to tell the truth? If they had, no one would have wanted to move out west. The romanticized imagery worked even though Runnymede did not survive.

This same romanticized fascination for cowboys and Indians that drew these earlier settlers to the Wild West continues today for many tourists. Western European tourists want to visit "dude ranches". Their image of an Indian is a person who wears the outfits that urban myths suggest that they should wear, e.g., buckskin clothing, long hair that is braided, and an eagle feather hooked through your single braid. European tourists today want to visit Indian reservations when on vacation in the U.S. They want to spend time at western dude ranches to practice "playing" cowboys and Indians. These experiences, however, can be similar to going to a zoo for those living on the reservation, e.g., when tourists arrive, they want to have their picture taken with a "real Indian".

Dude ranches are similar to visiting Disneyland except the tourist has the pleasure of roping a tame cow or riding a horse while wearing cowboy boots, spurs, a leather vest and preferably a 10 gallon cowboy hat. A decidedly sanitized movie experience! This experience is similar to the French courtesans during Louis XIV's time dressing up in peasant clothing and playing out the "idealic" good life of a shepherdess. They really had no grasp of how hard this life really was. For the French courtesans it was entertainment and they really were not interested in experiencing the hard life of a real shepherdess. It allowed them for a few hours or days to relax away from court politics back in Paris.

This fascination with the old west continues unabated today. A wealthy American businessman is turning his fascination with the west of the past by building a "faux" town to house his collection of western memorabilia. He is investing millions of dollars in building this western town on his lands in Colorado. This town is being built only for his pleasure and that of his friends. It is not a "faux" town that will generate income from paying tourists like many of the western ghost towns. This faux town is described in a newspaper article written in the Huffington Post [89]:

"There's a new town in Colorado. It has about 50 buildings, including a saloon, a church, a jail, a firehouse, a livery and a train station. Soon, it will have a mansion on a hill so the town's founder can look

> down on his creation. But don't expect to move here—or even to visit.
>
> This town is billionaire Bill Koch's fascination with the Old West rendered in bricks and mortar. It sits on a 420-acre meadow on his Bear Ranch below the Raggeds Wilderness Area in Gunnison County. It's an unpopulated, faux Western town that might boggle the mind of anyone who ever had a playhouse. Its full-size buildings come with polished brass and carved-mahogany details and are fronted with board sidewalks and underpinned by a water-treatment system. A locked gate with guards screens who comes and goes.
>
> Koch's project manager has told county officials that the enclave in the middle of the 6,400-acre Bear Ranch won't ever be open to the public. It is simply for Koch's amusement and for that of his family and friends—and historians. It is the ultimate repository for his huge collection of Western memorabilia."

Interestingly, in contrast to the western industrialized view, the indigenous communities in Indonesia do not want to mimic the Native American. They do not romanticize what it means to be an American Indian. They want to learn about Native American governance structure and how they make economic and social decisions. They do not have a cosmetic or artificial view of Native Americans. They feel a bond with them because of their cultural and traditional knowledge and values. They are interested in learning from the American tribal successes, i.e., lessons learnt in resources management or governance structures. Most global indigenous communities continue to have strong traditional and cultural roots that they want to maintain or restore despite living within the confines and constraints of a highly industrializing country.

### 6.2.2 Stories of "Real" Indians

Our contention is that people following the western European model of human development need to become "Indian" if we are to humanize their decision process. To become an Indian cannot be a cosmetic choice or where the facsimile is propped up as success. This will be a challenging endeavor since the western industrialized model of the "perfect man" [90] is not an Indian.

The western world listens to those who satisfy their idea of what they need to look like and how they need to behave. You trust someone who looks like and acts like you. A Wall Street Journal article summarized what is the "perfect man" based on the readers of E-book readers who could customize their romance novels [90]. "The perfect man, according to data collected by digital publisher Coliloquy from romance-novel readers, has a European accent and is in his 30s with black hair and green eyes. Optimal chest-hair level: "slightly hairy" [90]. Forty-two percent of the readers also preferred a "rugged body type".

You might say that this has nothing to do with how we make decisions but it does. There is a large data based suggesting that our preferences and the decisions that we make are influenced by who we trust and whose information we consider to be valid. We also like people to be similar to us, e.g., if company boss plays golf or tennis, it behooves an employee wanting to move up in a company to also take up these sports.

The other challenge to becoming Indian is that there are many biases and perceptions that have become rooted in society of what it means to be an Indian. These biases do not reflect the reality of how enterprising and adaptive Native Americans are. A brief family history of two tribal leaders follows to highlight how they do not match any of our views of Native Americans.

The photo of Mike Marchand characterizes well how today's Indians do not look like what the western European world has fabricated in their minds (Fig. 6.2; Mike Marchand, member and former leader of the Colville Confederated Tribes; Photo taken in 2012). He is outfitted in motorcycle gear and would have no problem fitting into the western European societies who enjoy riding their motorcycles. But he is also a tribal leader. Here is his story.

**Fig. 6.2** Mike Marchand.
Photo courtesy: Marchand.

#### 6.2.2.1 The Life of Mike as Told by Mike[1]

I am a member of the Colville Confederated Tribes in Washington State. I was born and raised on the Colville Indian Reservation. My mother, Thelma Cleveland Marchand, was a Councilperson for the Colvilles and my maternal grandfather was the tribal Chairman, John Cleveland. They were Wenatchees, one of the 12 tribes placed on Colville.

1 Confederated Tribes of the Colville.

The Wenatchee people had been promised by treaty a 9.7 km by 9.7 km Wenatchee Reservation at the present town site of Leavenworth, which is now famous as a faux Bavarian Village popular with tourists. My father, Edward Marchand, was the grandson of the last hereditary Chief Aurapahkin, of the Arrow Lakes people, who also ended up on the Colville Reservation. My dad was an airplane pilot and trained kids to enter into the construction trades. He liked to build and also was in charge of facility management for 0.6 million hectares of land. I grew up hunting and fishing and gathering foods with my extended families, much as our people have done for thousands of years.

My grandmother Mary Marchand preferred the traditional lifestyle. She followed the fishing and berries and hunting around throughout the year. She preferred living in a tent and camping out under the stars. So I spent a lot of time playing on the rivers and lakes and mountains. She was actually relatively wealthy and had three houses, so living in tents was a lifestyle choice.

Both my grandfathers were entrepreneurs. William Marchand ran a ranch, raised champion Hereford cows, operated a small sawmill and supplied ties to the railroad, made boards for houses in the area, and he also worked in silver mines. My other grandfather, John Cleveland, ran a big ranch, raised alfalfa and loved to race horses. Both ranches were very self sufficient, with big gardens and fruit trees.

At a young age, the Bureau of Indian Affairs noticed I was getting high 99 percentile test scores and picked me out to go to boarding school (see Box 17). They asked me where I wanted to attend school. I jokingly said send me to the best one in the world. A year later the government came back and handed me tickets to Philips Exeter Academy in New Hampshire.

**Box 17: Mike's Letter to Santa Claus**

The short clip from the Omak Chronicle published Mike Marchand's letter to Santa Claus when he was 9-years-old. In his letter, he asked Santa Claus for books and books on Leonardo Da Vinci and Archimedes.

Already at the age of 9 years, Mike did not follow the western world myth of what the behavior of a child who lives on a reservation.

Most 9-year-olds, even non-tribal people living off reservation, are not asking Santa Claus for books written about these western world thinkers and artists.

☆ ☆ ☆

**Dear Sanata,**
I want a watch. I want two shirts and two pair of jeans. I want a Jimmy Jet. I want some perfume. **Kenny Kay Vaughan &**

☆ ☆ ☆

**Dear Santa,**
I want a biology set and book and books about Leanardo Da Vincy and Archimedes. I want a leather craft set, a dart game, a couple mosaic tile kits.
**Michael Marchand 9**

☆ ☆ ☆

**Dear Santa,**
I'd like to be a Scientist when I grow up. So I'd like to have book's on science and rock's. I know if I may have them I will be happy.
P. S. I wish for piece on Earth. **Peggy Sue Wilson, 9**

☆ ☆ ☆

So I went there and was fascinated with French class and Tudor Stuart English History of all things. While there a life changing experience happened. Some friends came by one weekend; it was Thanksgiving weekend of 1970. I thought we were going to some Thanksgiving Dinner. It turns out we were going to one of the first big Native American protest demonstrations at Plymouth Rock. Dennis Banks and Russel Means of the American Indian Movement were the leaders. Russel was giving some great speeches, fire from the belly type speeches, about how we need to be sovereign, act sovereign, and get the Bureau of Indian Affairs out of the way. While listening to him I was thinking OK, who is going to be planning and building this new Indian world. I thought I could help do that. So, I went on and studied urban and regional planning and studied business development and economics. I left Exeter and went home, the east coast seemed too distant and irrelevant to my life after that Thanksgiving Day.

In 1970, there was virtually nothing at the Colville Tribe in terms of development. Due to the termination policies, there was only a skeleton crew at the Agency, less than 10 employees. But due to the strong hearts and determination of the Colville people, with help from Indians across the nation, the termination of the Colville Reservation was stopped. The liquidation of billions of dollars in assets by the US was halted. The Klamath Tribe in Oregon was not so lucky. Their lands were terminated. Our tribe was next on the chopping block. But we were able to save ourselves.

I obtained a Bachelor of Arts degree from Eastern Washington University, majoring in Economics and Urban and Regional Planning. I also received a Master in Urban and Regional Planning from Eastern Washington University. Currently I just finished my PhD in the School of Environmental and Forest Sciences, College of the Environment at the University of Washington.

Another Indian who is an important western world and Native American leader is John McCoy. His story is given next as well as a recent photo of his (Fig. 6.3).

**Fig. 6.3** John McCoy.
Photo courtesy: John McCoy.

### 6.2.3 The Life of John McCoy[1]

Even though my father had joined the Navy before I was born, I was born on the Tulalip reservation. My father served 30 years in the Navy but would spend 30 days of leave each year on the reservation. So I was mainly raised on the west coast since my father's Navy career had him mostly stationed in San Diego, California. When I was in the 2nd grade, my family moved back to the reservation to live so I lived both in San Diego and the reservation.

I graduated from high school in San Diego and then returned to the Tulalip reservation and fished for five months. However fishing requires you to get up at 3 am in the morning and work until 10 pm at night. I decided that there were other jobs that were more suitable for my interests than fishing.

I had always been interested in computers. So I joined the Air Force. After 20 years, I retired in 1981. I had accumulated a great deal of training in computer operations and programming. The explosion of computers in the mid-1960s meant that when I retired, I did not have to look for a job and had six companies recruiting me. For my first job, the company I joined had a contract with the White House to automate its computer network. I worked as a computer technician in the White House from 1982 to 1985. I started as a "grunt" when I first started this job but eventually became the site manager.

Starting in 1992, the economic downturn meant that the company I was working for had to lay off people. Because of my position, I was the person who had to tell people who had become my best friends that they no longer had a job. Again in 1993, I had to lay off more people. Since I did not like this part of the job, I decided to approach the Chair of the Tulalip Tribes to see if the job he had tried to recruit me for in 1992 was still available. This job was the economic development manager for the Tulalip Tribes. Fortunately for me, the job was still available. In 1994 I returned to Tulalip to help bring the community into the digital world and built what is now the Quil Ceda Village Business. In 2000 I became General Manager of the Quil Ceda Village Business Park. After 11 years, I retired from this job in 2011. The story of how I helped Tulalip Tribes develop some of their business enterprises continues later in our book (see section 10.3.1).

My oldest daughter first became involved in politics and worked for Morris Udall for five years. I followed my daughter into politics. I became a state and national leader on diverse, important issues involving broadband, alternative energy and K-12 education. I am also very active with the Native American advisory boards for the National and Snohomish County Boys and Girls clubs.

I represent the 38th Legislative District, which includes the Everett, Marysville and Tulalip communities of Snohomish County. I am also the Chair of the

1 As recounted by John, Tulalip Tribe.

National Caucus of Native American State Legislators. I am an active member of the Environmental Management Roundtable, the Labor and Economic Development Committee, and the Communications, Financial Services & Interstate Commerce Committee for the National Conference of State Legislatures. In the Washington State Legislature, I was the former Chair of the House Technology, Energy, & Communication (TEC) Committee and sit on the House Education Committee and the House State Government and Tribal Affairs Committee.

I have worked on expanding broadband across the state of Washington to get folks connected. I have also pushed to expand Washington's renewable energy portfolio and create incentives for alternative energy, like biomass and solar. I continue to work on making Washington State greener and run more efficiently. In addition to the environment, energy and development issues occurring at the state level, I work on energy issues that impact tribes as well as the Nation. I am a member of the National Environment Management Roundtable on nuclear spent fuel and also a member of the State Tribal Governmental Working Group on nuclear site cleanup within the U.S. Department of Energy.

A characteristic of both tribal leaders is their ability to bridge and work towards consensus on tricky problems. Both have been successful in developing policies that are compatible with tribes and state or federal governments. An important message is that Native Americans do not sit in their reservations and ignore the world that exists outside of these bordered lands. They also think about the impact of their actions on future generations of tribal and non-tribal peoples. They have incorporated the elements of the portfolio for practicing sustainability that has been written in this book. They are perfect examples of tribal leaders who have successfully blended Native American behavior with the western science approach. They are good role models that our portfolio is practical and works!!

## 6.3 Think Slowly and for the 7th Generation

*This was a merchant who sold pills that had been invented to quench thirst. "You need only swallow one pill a week, and you would feel no need of anything to drink."*

*"Why are you selling those?" asked the little prince.*

*"Because they save a tremendous amount of time," said the merchant. "Computations have been made by experts. With these pills, you save fifty-three minutes in every week."*

*"And what do I do with those fifty-three minutes?"*
*"Anything you like..."*

*"As for me," said the little prince to himself, "if I had fifty-three minutes to spend as I liked, I should walk at my leisure toward a spring of fresh water."*

~ Antoine de Saint-Exupéry, *The Little Prince* [91] ~

The excerpt from *The Little Prince* exemplifies the choices that society faces daily. Technology has made it where each one of us can experience nature synthetically and very quickly. This synthetic solution is not acceptable to a Native American. Most indigenous communities will not abandon their pleasures of experiencing life or nature at nature's time and spatial scales. Many in the western world take the pill so that one does not waste time in searching for food or water or to have an exhilarating experience. This is an easy decision to make since western world citizens are disconnected from their environment. They are clueless as to where resources are produced and what is needed for nature to continue to deliver ecosystem services.

The tribes have a different concept of time and what you do with your time than Europeans (see section 8.3). This view is expressed well in the following Taoist proverb:

> No one can see their reflection in running water. It is only in still water that we can see.

The use of technology has a cultural and ethical consideration in its use that transcends across several generations. One of the authors (John D Tovey) described it this way:

> My idea of Seven Generations may not be a literal 7 generations, but the time something remains in one's collective memory. I can know of someone 3-4 generations ago (I know my grandmother personally who knew her grandmother personally for a total of 6 generations). This was not second hand connections. The 7th is beyond that knowledge but we must acknowledge that it is there, and there will be a 7th generation that will not know us (remember us). This connection means that it will affect our decisions.

Europeans use a linear timeline, which is foreign to most tribes. A 20-year business plan projection is considered long to the typical European or their descendants, but for the tribes 50+ years is more important. Tribes consider anything less to be short-term.

Tribes know which way they are moving but are not as worried about what the end-point is. This could be compared to a "systems thinker" versus a "linear thinker" (see section 3.2.1). When one is mostly a linear thinker, an end-point is identified and one works studiously toward reaching it. This focus places "blinders" on a person. It limits their ability to recognize when the end-point is out of reach or when one needs to adapt to new facts. If technology controls one's approach, it is easier to be a linear thinker and to accept that technology will ensure that the right choice was made.

Thinking like a system and being adaptive are extremely hard for most people to practice. You have to accept that you do not have complete knowledge of any problem and will never know enough to make a final decision. Or, there is not enough time to contemplate alternative decisions in the shortness of the decision time.

Tribes also accept that we do not know all the answers, so the decision process is taken more as small steps. Western European perspectives do not explicitly acknowledge that science is incomplete. Instead expert opinion and sophisticated models are used to support assumptions and the decisions made.

Since models are a summarization of our knowledge, it can only be as good as the current knowledge. Despite each model being built using thousands of data points, it will not produce a better decision if data outliers are used or they do not represent the norm. Since only data that are correlated to one another are used, it propagates a linear thinking approach and supports that pre-identified end-points are valid. Western Europeans also recognize that we have accumulated volumes of knowledge in libraries so the foundations of our decisions are based on a solid foundation.

The urgency of decision-making triggers short-term decisions that may misdiagnose the real problem. A recent Wall Street Journal article is revealing to read since it highlights the short attention span of today's western world society (Box 18).

By contrast, the tribal decision process is long-term and protracted. Tribes make decisions for the 7th generation and would not make decisions that decrease their ability to adapt to their environment. Western Europeans make decisions for the short term and allow these decisions to drive politics and economics. Tribes make locally based decisions informed by the larger landscape and elders while western world scientists use science generated in other geographic locations to make local decisions.

Western Europeans need to develop a culture that is applicable to a local environment but informed by science collected at distant locations. If they do not change their decision process, the western world will be like weeds that appear to thrive locally but are converting nature into a novel ecosystem. Since

**Box 18: Western World's story of Time**

"Today's signature move is the head swivel. It is the age of look-then-look-away. Our average attention span halved in a decade, from 12 to five minutes, according to a study commissioned by Lloyds TSB Insurance. (And that was in 2008.) We miss almost everything; we text while we walk. What makes a person stand out now is the ability to look and keep looking.

A "museum intervention" is now mandatory at Yale's School of Medicine for all first-year medical students. Called Enhancing Observational Skills, the program asks students to look at and then describe paintings—not Pollocks and Picassos but Victorian pieces, with whole people in them. The aim? To improve diagnostic knack."

(from Finn [92])

weedy species can dominate in the short term, it may look like everything is okay but these species will be replaced when humans no longer manage an area.

## 6.4 Long Scientific History but Short Human Memory

The western European model of credible knowledge is derived and disseminated by University-trained researchers and scientists. They collect, research, test, generate hypotheses, and finally filter what knowledge is acceptable and will be used to make decisions. Universities are the repositories of knowledge and educate the next generation of decision-makers. University educational systems are the core 'dispersal mechanism' for knowledge and the repository of 'credible' science in the western world.

A University-developed knowledge base by definition should have strong historical roots. The first University was built more than a thousand years ago. According to the Guinness Book of World Records, the University of Al-Karaouine in Morocco was founded in 859 CE by a woman. Despite this long history and the countless books written since that time, scientific knowledge in practice may go back a couple of decades, if even that far back in time. It is common knowledge among University professors that a recently published research paper is cited for about five years. Once this time has passed, a science paper is frequently lost from the scientific memory and is no longer referenced to by new publications. This happens despite the fact that we have libraries as repositories of books and digital technology that can 'mine' data and provide information at our finger tips.

The unfortunate repercussion of this short-term memory of past scientific research is that past scientific paradigms are rediscovered in the present. These past papers are not referenced for the ideas that they generated but are introduced as new paradigms. This means that's that science repeats itself and scientific knowledge recycles periodically, i.e., it is rediscovered many times. Leading scientists have talked about scavenging old library journals or books to look for new insights for today's pressing problems. These ideas are then reintroduced as new and novel ideas—a great way to build a scientific reputation since the idea has already been tested so failure is unlikely. This approach is justified since today's tools for measuring this phenomenon are more sensitive and able to measure the finer details of any process. However, the core idea has not fundamentally changed.

Even young adults today have less interest in history and begin their world view starting when they were born. While teaching a class two years ago at a premier research university, one of the authors (Kristiina A Vogt) of this book had a student ask why they needed to know some fact that had occurred before 1978. His view was that only relevant information occurred after his birth date. Since he was born in 1978, his history started on that date. He had no historical vision and really did not see relevance for anything that happened before he became an active 'player' on this world's stage. He asked this question without recognizing that he is what he is because of his family history and beliefs that were probably rooted in stories written a thousand years earlier.

In the Introduction to the "The Historian's Craft", written by Marc Block and published posthumously in 1953, Strayer poses the same question about the relevance of history as he introduces the importance of historians [93].

> "... What is the use of history?" What is the use of history, when the values of the past are being ruthlessly discarded? What is the use of history, when we repeat our old errors over and over again?..."

Using a coin as an analogy, some see science as being the other face of economics or the other side of the coin. In 1953, Block wrote [93]

> "it is undeniable that a science will always seem to us somehow incomplete if it cannot, sooner or later, in one way or another, aid us to live better." He continues, however, ... history is not only a science in movement. ... but... it struggles to penetrate beneath the mere surface of actions, rejecting not only the temptations of legend and rhetoric, but the still more dangerous modern poisons of routine learning and empiricism parading as common sense."

When searching for solutions, tribal members consult tribal elders to gather historical knowledge passed down inter-generationally. As John D Tovey noted (see section 4.5), *tribal elders are equivalent to the Google Search Engines of pre-Columbian knowledge.* In the western world, historians are our reservoirs of knowledge. But today, historians are mostly asked to comment on *past actions* of important people or events that occurred. It is not uncommon to hear that history will describe someone better in the future than what is possible today. More facts will be revealed as historians are able to study documents from the time. Therefore University professors have great difficulty in adequately responding to a question of how our current decision will impact another problem. We do not have enough history to analyze this other problem. Even history is hard to read for insights because much of the information is fragmentary and winners of battles always write their own history. Any rulers history will always be flattering and not mention the bad decisions they may have made.

When institutions of higher education are the gate keepers of knowledge, they can also control what and when knowledge is revealed. Since educators decide what knowledge is taught at schools and Universities, they influence the frameworks and approaches decision-makers think about using to conduct an assessment. They also control what information or facts becomes acceptable knowledge because they are able to either reject or accept papers that they review for publication. They select who will receive funding for their research.

In the mid-1970s, one of the authors of this book (Kristiina A Vogt) witnessed how the scientific leaders in a particular discipline blocked attempts by other scientists to publish their ideas since it was counter to what had become the acceptable pillars of knowledge. In this case, these scientists were the leading researchers in their field and had been instrumental in developing what was accepted as the foundational research. Anyone contradicting their research was unable to publish their work. Fortunately this situation is not the norm. But it does happen. It tends to be easy for those identified as the leaders to be able to stifle views counter to theirs.

If history was integrated into our current knowledge, it would reveal the evolving ideas that are the pillars of science. It would be harder for scientists and professional to address complex sustainability problems using a too narrow perspective. This narrow perspective can be broadened by adopting a historical approach or the decision processes used by Native Americans as part of integrative decision frameworks.

The Native Americans, as well as most indigenous communities, do not have the luxury of moving elsewhere if they make bad decisions on their lands. They have to live with their mistakes and accept the negative repercussions of their decisions on the community. The entire community will have to absorb and deal with these individual mistakes. This is not a common occurrence in Native American communities since cultural traditions provide feedback that minimizes many bad decisions being made. Since the feedback is immediate to the individual when he/she makes a bad decision for the community, fewer bad decisions will be made. Course correction is possible because there is not a lag time between an action taken and feedback made on a bad decision. This feedback is part of the cultural traditions.

By contrast, a bad decision made by an individual in the western world is not absorbed by the community. The community will not take responsibility for an individual's error. Even the individual may not have to take responsibility for his or her bad decision. If an individual is linked to a bad decision, he/she may be forced to resign but no longer remains in the company to deal with the bad decision.

There is also a long lag period between when a decision is made and when there is recognition that the decision was wrong. This long lag period may also make it difficult to identify a specific action that caused a bad decision. In this case, no one has to take responsibility for the bad decision even though the group will still have to be dealt with its repercussions. Short-term course corrections are also difficult to make due to a lack of historical knowledge at the ecosystem level and how the parts are connected. People living in cities are especially isolated from the repercussions of the decisions they make. Their group survival is not dependent on individual decisions.

Policy-makers rely on scientists to provide them the historical memory or to place any science in its appropriate context. Policy-makers, however, do not want the historical memory when it slows down the decision-process or when it does not support their political views. Depending on your values, science can become a biased venture where only science that supports one's views is presented. This has been especially problematic in the natural resource sciences since competing demands for scarce ecosystem services make it hard to decide which resources should be given priority. In such cases, the easiest solution is to follow your values since it is more comfortable to make decisions framed by what you deem to be acceptable science.

Since human values can drive what information is used to make decisions, individual biases become embedded in the decision process. If one fears forests or large mammals, it will be easy to agree to cut a forest or to kill an animal. Hylophobia or fear of forests is found among many of today's American public [94] so this is not a farfetched idea.

A survey of Native Americans who live on a reservation or return frequently home would find few that are afraid of nature and/or forests. Their livelihoods and cultural values mean that they are integrally linked to nature. They are not afraid of the environment that exists outside of the "human built environment".

Native Americans have adapted to the western world but retained the inter-generational knowledge essential for making natural resource decisions. John D Tovey recounts how his tribe was concerned with the loss of inter-generational knowledge after many tribal members no longer lived on the reservation. In this story his grandmother's generation was important for maintaining inter-generational cultures and traditions.

## 6.5 Inter-generational Adaptation and Grandmothers as Told by John D Tovey[1]

My grandmother is 3 quarters Cayuse and 1 quarter Joseph Band Nez Perce. I grew up nearly 800 kilometers away from the Umatilla Indian Reservation, and only visited a handful of times for funerals until my father (her son) became employed at the reservation when I was about 9 years old. For much of my life I was surrounded by the other 3 quarters of my heritage: Welsh descendent Mormon immigrants who settled in southeastern Idaho. The only real window I had into the other 1 quarter of me was my grandmother. My father and his brothers, to no fault of their own, were part of the "Denial" generation of Indian descendants when "being Indian" was associated with the results of 100 years of post-treaty colonization, i.e. poverty, crime, alcoholism, welfare, etc. This "Denial" generation, in my estimation, later became the "Greatest" Generation.

My grandmother told me stories about being a child on the reservation and how many people would discourage their youth to leave for educations, marriage, or other opportunities. These elders, as children, had seen their fathers, mothers, siblings and friends die on battlefields, collapse from exhaustion during forced marches, and were brought up at near starvation on mixtures of government-provided taro flour, baking soda, water and salt; now famous as Indian Frybread. These elders saw their way of life destroyed, their lands lost, and their language die. They didn't want their children and grandchildren to leave for school not because they wanted company in their misery; rather they felt that all they had left were their children. To allow them to leave, in their estimation, would be the final nail in their cultural coffin, because why on the Great Spirit's Earth would these children ever want to come back to this deplorable mess? A few wise elders, however, realized that their children needed to learn, adapt and become part of this greater conquering culture *in order* to survive. In my grandmother's estimation, her parents were of this later type.

And so many of my grandmother's generation left, and more of the following generation left. It seemed the fears of the protectionist elders were coming true until something amazing started to happen. These children and especially the grandchildren (The Denial Generation) who had gone off become lawyers, accountants, doctors, nurses (like my grandmother) and business men (like my

1 Cayuse, Joseph Band Nez Perce.

father and uncles) and after having successful careers, they began coming home to care for elderly family members or just to be close to loved ones. They brought with them a deeper understanding of the external world and with the lens of their heritage began utilizing this knowledge for the betterment of their community. A few lawyers and businessmen realized the full applicability of the Commerce Clause of the standard treaties, and this leads to the explosive economic development enjoyed and explored by many of the tribes today. This new economic awakening has led to other awakenings of culture, experience, health and human welfare, and natural resource management that could all be reasserted by utilizing western knowledge *through* a cultural lens. This lens, I believe, will be one of the greatest contributions to management of natural resources indigenous people will make for the future of our planet and race.

For the last 200–300 years western thought has asserted that the management of natural resources would be based on economic development (the purely western notion anyway), arguably much to the environment's degradation. But with a traditional and cultural lens applied we can no longer look at natural resource services as isolated economic elements to be maximized and exploited. Granted this is not a novel understanding of natural resource management. Complexity theory and human-coupled models of natural resource management particularly with relation to human understanding have been gaining more and more of a foot held for nearly 50 years. But there is still something missing from this academic jargon of ever increasingly complex interlaced systems. This is where native notions of natural resource management become relevant.

We have several lenses that need to be combined to become adaptive ecosystem managers. This is practiced by the Iban Tribe in West Kalimantan in Indonesia. Their story is recounted next.

## 6.6 Cultural Diversity the Norm in Regional Landscapes: Iban Tribe, Indonesian Borneo

Iban tribe is one of many different Indigenous groups of Borneo. They live in northwest part of Kalimantan (Indonesian Borneo) and the southwest of Malaysian Borneo. Indigenous groups of Indonesian Borneo have been homogenized as the Dayak tribe. The term Dayak actually represents the term that refers to different tribes and sub-tribes in Kalimantan. This term was coined during the Dutch colonial administration to distinguish those ethnic groups from the Moslem ethnic groups who mostly lived in coastal areas of Kalimantan. According to one of Dayak activists John Bamba, the Dayak people initially did not identify themselves with this term as they called themselves by the names of their own tribes and sub-tribes (noted in an online bulletin Perspektif Baru, Edition 601, September 2007). Hundreds of Dayak tribes and sub-tribes have their own names such as the Iban, Kayan, Kenyah, Jalai, Kanayan, and so on. See Selato [95] for detailed examination of the Dayak.

One of the homelands of the Iban People in Kalimantan is the Sei Utik customary area, which is a customary village located in Embaloh Hulu sub-district, Kapuas Hulu district, West Kalimantan Province, Indonesia. Sei Utik borders with Sarawak in the north, with East Kalimantan Province in the east, and with Sintang district of West Kalimantan Province in the west. Sei Utik is part of the Jalai Lintang customary area (ketemenggungan). Other parts of this customary area are customary villages namely, Lauk Rugun, Mungguk, Pulan, Kulan, Apan (Langgan Baru), Ungak, and Sungai Tebelian. According to the government administration system, these villages are incorporated into the village of River Utik (Menua Sei Utik).

In Sei Utik, the Ibans live in a wooden long house (see Fig. 6.4). This long house has four main parts. The first part is the individual house that looks like an apartment unit. There are around 25 units of individual houses in this long

**Fig. 6.4** Long House (top left), The Communal Hall (top right), Community gathering at the Communal Hall (bottom left) and Communal Porch of Iban Tribe in Sei Utik, West Kalimantan, Indonesia (bottom right).

Photos: Mia Siscawati 2010.

house. The second part of the long house is the long hall located in front of the front doors of the individual houses. This long hall is a communal area where the Ibans of Sei Utik do individual, familial and communal activities such as weaving baskets, storytelling, town hall meeting, gathering.

Certain parts of this long hall are also used to put tall baskets that are full of non-timber forest products. The third part of the long house is the long porch, which is usually used for gathering of certain groups such as women and children, and for keeping agricultural tools and other tools. The fourth part of the long house is a communal "yard" that is usually used to dry paddy from dry fields and other products from their swidden agriculture.

Most Iban's households in Sei Utik make a living from a variety of activities where natural resources play critical roles. Swidden agriculture activities—including dry-field (umai pantai) and wet-rice (umai payak) farming as well as rubber garden and mixed trees gardens—serve as sources of food, traditional medicines, daily cash (especially from tapping and selling their rubber trees), and a combination of mid-term and long-term savings (from annual harvest of tembawang, rattan, and fruits). Sei Utik community maintains customary rituals that connect the relationships between people, e.g., covering birth, marriage, and death as well as the relationship between people and nature.

Sei Utik has served as a place where a number of NGOs, activists, and progressive academicians learn about the application of community-based forest system (SHK), including how this system develops and operates under customary laws. The learning process also includes the topics of customary spatial arrangement and complex tenure system, covering a variety of concepts of tree tenure and land tenure. Besides serving as a learning site on SHK, Sei Utik has also served as a site of collaborative community organizings where NGOs and Indigenous Peoples Organizations (IPOs) have been working together with the Sei Utik community to address external pressures that threaten Sei Utik's forests and the whole of Sei Utik's customary territory. The external pressures include illegal logging, financed by Malaysian entrepreneurs from across the border, and oil palm plantations planned under the Indonesian-Malaysian border mega-project. These have the potential to devastate the people of Sei Utik's forests and livelihoods [96]. According to Pak Janggut (Chairman of the Iban Tribe in Sei Utik), "the biggest problem is to prevent the forest area from being converted into Industrial Plantation Forest area or Palm oil plantation area."

Responding to the above external threats, the Iban community in Sei Utik, together with several local groups (Program Pemberdayaan Sistem Hutan Kerakyatan/PPSHK, Lembaga Bela Banua Talino/LBBT, Pancur Kasih) has developed various initiatives and alternatives to defend the forests. The initiatives developed in Sei Utik with these supporting groups include [96]:

- A Credit Union developed with Pancur Kasih to strengthen the local economy and reduce internal pressures on the customary forests;

- An initiative developed with LBBT to build and strengthen the political position of the Sei Utik Iban community. A study to identify their customary/ancestral rights (hak ulayat) has been done. This has provided material for drafting a Perda (local government regulation) which recognizes the existence of the Sei Utik community and their customary area; and
- Participatory community mapping process and participatory planning process of the Sei Utik's forests facilitated by PPSHK Kalbar.

The Ibans use cultural values to decide what and how to use resources. Certain tree species are used as construction materials to build boats and fire-wood despite the fact that these tree species are high valued and easily sold in international markets. The Sei Utik forest has an abundance and diversity of timber species so timber scarcity is not part of the decision process. The Sei Utik forests are rich in biodiversity and provide the community with timber as well as many non-timber products. The dominant tree species are meranti (*Shorea* spp.) and kapur (*Dryobalanops* spp.). Other tree species found in this forest are ladan, gerunggang (*Cratoxylum* spp.), kempas (*Koompassia* spp.), and jelutung (*Dyera* spp.).

The total forest area managed by customary community of Sei Utik is 9452.5 ha. This area is divided into four categories of land-use:
- Kampong Taroh (Protected Forest) covering 3,667.2 ha;
- Kampong Galao (Forest Reserves) covering 1,510.7 ha;
- Kampong Embor Kerja (production forest) covering 1,596 ha; and
- The remaining land area is managed for agriculture and other purposes (2,678.5 ha).

The entire allotment of forest land is under the customary rules of the Iban of Sei Utik village. The Sungai Utik customary forest area has been "successful" in acquiring several forms of recognition for how well they have managed their resources. One form of recognition was the certification label for their "sustainably managed forest" through the Indonesian Ecolabelling Institute's special certification scheme for customary forests. The external pressures that Apai Janggut shared with one of the authors of this book (MS) in 1999 were still threatening Sei Utik at that time.

**Coyote Essentials**

- Consider every component in the ecosystem and ecoregion when making land and resource management decisions.

- Recognize the biases you have and how they influence your decisions.
- Just because an old fish is wrapped in decorative paper doesn't mean it won't be smelly and bad to eat.
- Slow down and think for the 7th generation so you do not misdiagnose nature problems.
- Do not ignore science older than the day you were born.
- Really talk to your grandmother and grandfather, and learn from them.

# Chapter 7

# Portfolio Element III: Follow a Native American Business Model

*Only after the last tree has been cut down,*
*Only after the last river has been poisoned,*
*Only after the last fish has been caught,*
*Only then will you find money cannot be eaten.*

~ Cree Prophecy - //www.stevenredhead.com/Native/ ~

## 7.1 Company Business Plans or Village Economics

There are a number of differences between a Native American run enterprise and European style undertaking:

- All Native American business plans are designed to take into consideration the life and livelihood of the 7th generations beyond those individuals making the decisions;
- Since most of the undertakings are within the boundaries of tribal lands, they will live with the decisions they make, both good and bad, on a daily basis;
- Conserving their way of life and the environment for the future is more important than having financial gain now.

Native Americans don't produce a 'sales pitch' to investors and expect them to not only to provide all the financing but to totally shoulder the risks alone. The tribe shoulders the risks with any business venture that they decide to develop.

When economics is the only factor that determines how societies interconnect with nature, controlling and dominating nature is a very well developed practice. Historically economics has been the dominant western world tool to make decisions on what, how and where to collect natural resources. For a business to increase its economic returns from nature, you need to own nature and be able to build walls around your resource. This means that our sense of property rights and ownership is well developed since we need to ensure that the revenue returns to the "owners" of the resource. If you own a piece of nature, you will also feel that you have a right to do whatever you want to with your piece of nature.

Native Americans do not own nature in contrast to the western world view where lands are bought and passed down through inheritances to a successive list of owners. Nature is communally owned and not owned by an individual Native American.

It is worthwhile exploring in greater detail the differences between non-tribal and tribal business plans since this is a fundamental difference between American tribes and western world businesses.

### 7.1.1 Non-tribal Business Plans

The majority of the companies in the world today first develop a business plan when they start a new economic venture or rebuild an existing business. The business plan lets the managers know whether their venture is too costly or where the revenues would balance the costs. It could have been written on the back of an envelope or written by an MBA working for a firm that specializes in this type of work. Perhaps the business plan came from the numerous sites available on the internet that offers the form of pre-packaged business plans. The latter approach provides a generic or one size fits all business plans.

Business plans are complex and contain a considerable amount of detailed information. It typically describes the company, its founder, and what products the industry will produce. It has to address who are the competitors who produce or sell a similar product, what are the future trends for people wanting to consume this product or what will be the future product demand. Furthermore, all this information has to describe where the company presently sits within that industry and how the company will meet those goals. In addition, this business plan usually provides a company balance sheet with financial details and the internal rate of return if a new piece of machinery is to be manufactured. It also includes a management structure and rules for how decisions will be made. It also provides what are the responsibilities of all executives in the company.

A business plan is defined as an essential road map for a company to evaluate its success potential. It also identifies the potential risks that could derail this business venture. A business plan usually is a 3- to 5-year plan that outlines what avenue the company intends to follow to grow its revenue. This is a living document that changes depending on the circumstances and outside forces that the company has no control over. Most business plans are developed when a company wants to request funds or loans from some type of financial lending institution or investor group.

As mentioned earlier a business plan is a living document that quite often is drastically changed by the time it is implemented. With the fast paced movement of global markets and communication networks today, in reality a Chief Executive Officer (CEO) is seldom able to plan for the long term. The responsibilities of a CEO are set by the Board of Directors of the organization. Some companies give CEOs broad latitude to run the company while others have more direct over-sight by the Board of Directors.

Ultimately, the CEO has to perform according to the expectations of the board and its stockowners if they expect to retain their position. If a company's quarterly profits are lower than what was expected, a CEO has to respond to the Board members and stockholders and pursue some sort of corrective action. Repeated lackluster business profits can cause a CEO to be fired from his or her position. CEOs being fired from their jobs are not an uncommon event and frequently reported in the business section of newspapers or business magazines.

To maintain their high paying jobs each company CEO must make certain that short-term profits satisfy the expectations of financial experts or the markets. This means that the company may be economically worse off in a few years. At least the short-term returns satisfy the market expectations for the company. Long-term strategies that may strengthen the company, but not provide dividends for stockholders, are rarely undertaken.

On some occasions stockholders can get involved and vote to remove members of the Board if they feel that the direction the company is not satisfactory. In addition, CEOs must keep in mind that stock analysts report their opinion on the value of a stock every quarter. Their attitudes regarding the value of a stock can have grave consequences for the dollar value at which the stock is sold on the stock market. It also means that short-term financial decisions made for the short-term or quarterly goals are the norm. Long-term goals are the exception.

The business decisions made by some CEOs are based on how long they plan to stay with a particular company before they move to a more financially rewarding situation. To get one of these better positions, they need to make certain that their present company looks good on paper no matter what problems their previous decisions have created for the near future of their present company. Whatever damage they have done to the company will have to be corrected by the next CEO who is hired to replace them.

Since CEOs do not have to live with the problems they create, they tend to not to be held accountable for their decisions. The recent trend for Board of Directors to reward CEOs, despite a company having a poor performance record, attests to the lack of accountability a CEO has for his or her poor decisions. Before beginning to work for any company, most CEOs have a contract written up that provides them no financial hardship if he/she financially hurts or bankrupts the company while they are in charge.

It is not easy to be a CEO of a company since there are so many unknown factors that impact whether a company will be profitable or not. There is a reason that business journals recount how the same dozen business leaders keep on being hired by different companies when business operations are unsatisfactory. It takes a creative and insightful business leader to improve the profitability of a company. Each company needs its own strategy for improving its profitability. This cannot be a 'cookie cutter' approach to conduct business. A CEO cannot follow the case studies that he or she memorized in Business School.

One of the dozen successful business leaders told one of us (Kristiina A Vogt) he was hired to fix a company that was not performing well. Other CEOs had failed to turn the company around. He had to significantly change the institutional structure and decision planning process to improve the profitability of this company. He had to change the entire company from having a vertical to a horizontal management structure. He mentioned how a vertical management structure isolated managers from knowing the information that would have improved the company's efficiency and therefore its finances. He recounted how in this company a train traveled to Seattle at a specific time and day. This schedule never varied. This situation occurred even if no product needed to be transported from Seattle to another location. This was a waste of company money and resources. He changed this. This CEO needed to understand and comprehend how all the parts of a complex and large company were connected. He could not focus on one part of the company. If he did, he would not have been successful.

### 7.1.2 Tribal Business Plans

The "business plan" of a Native American tribe is very different when they run a company or manage their tribe's resources. Using traditional knowledge focuses tribes on communal decision making. Resources are communal (see section 5.3.1) and therefore business enterprises have to satisfy the community. Furthermore since the production and consumer are not disconnected in the tribal view, simple solutions are not the norm. For example, tribes working on solutions to dams do not focus on the problems, particularly of the dam itself. This will not solve the original problems developed by the dam which is a lack of jobs and loss of traditional livelihood jobs, e.g., salmon fishing. The western Europeans did not factor in the economic impacts of building dams on the tribes but focused on the benefits they would get from pursuing these large water projects. For the tribes, they lost cultural resources and were excluded from being part of regional economic development goals. This is not how a Native American would practice their business mode. It also effectively removes tribes from practicing and building their brand of business enterprises that is culturally based.

Historically American tribes built successful regional business enterprises as collaboration among several tribal groups [8]. Tribes are not new to the art of managing a business.

Today, Native American business enterprises are emerging as some of the most successful regional economic engines in the US. They practice their own brand of business development and use the resources located on reservations to develop regional and global business enterprises. Miller [8] writes in his book entitled "Reservation Capitalism" how many American tribes have been spectacularly successful with their business ventures despite making business decisions that are not solely economically based. He also writes that some tribes have failed to develop their business ventures. Not all business ventures started by tribal people succeed just like many non-tribal businesses. What is noteworthy is the high number of successful business ventures that follow a tribal business model.

The many examples of successful business ventures found in the tribal communities are especially interesting considering what Native Americans have experienced (see section 2). Miller [8] describes well the diversities of business ventures in timber, minerals, land leasing, manufacturing, agriculture, ranching, and grazing; government administration; tourism; fishing; water; and housing, to name a few. These are some of the many examples of economic development opportunities pursued by multiple tribes.

Native Americans are not new business entrepreneurs. They built their business enterprises following the strong history of engagement in business ventures before the arrival of the western European colonialists. Miller [8] writes

> "...Some tribes today are among the leading employers in their states, such as the Mississippi Choctaw, or in their regions, like the Confederated Tribes of the Warm Springs Reservation in central Oregon, and the Confederated Tribes of the Umatilla Indian Reservation in northeastern Oregon..."

According to the Harvard Business School, successful tribal businesses have several attributes. The Harvard Project assessed over 100 tribal businesses enterprises to identify the key elements of successful tribal economic enterprises [8]:

- "First, tribal governments have to exercise their sovereignty... control and make their own decisions about what businesses to create and operate on reservations, how tribal natural resources... will be developed,... how businesses will be structured and what their missions will be...

- ...tribal governments ... develop strong institutions to assist and regulate business development...ensure the rule of law in Indian Country...tribal governments give people more procedural protections than do other governments...
- ...cultural issues are very important...Few tribal cultures and reservation populations...support businesses that are antithetical to their core beliefs and institutions..."

The Harvard Project identifies sovereignty and culture as key factors in successful tribal businesses. Culture is central to any tribal business venture. Potential business enterprises are not viable if a business activity cannot satisfy that there will be no impact on cultural resources located on a reservation. There are many examples of tribes *not* moving forward on a lucrative business venture because it posed a risk to cultural resources. Culture is not a negotiable item for any tribal business deal.

In addition to the factors already mentioned by the Harvard Project, several other metrics are important to satisfy before tribes are willing to agree to collaborate on or to move on a project. Colleen Cawston (former Chair of the Confederated Tribes of the Colville) summarized these factors eloquently to a University of Washington graduate class four years ago. Some factors she mentioned were also cited by the Harvard Project. The noteworthy comment is that the key characteristics of successful tribal businesses are not common elements of most successful western world business enterprises. Colleen mentioned:

- The importance of maintaining and restoring tribal languages—languages are the manner in which tribal members link to nature so any decision *cannot impact the survival and restoration of tribal languages* (see section 3.3.3). This was not specifically mentioned by the Harvard Project.
- *It is very important for tribes to make their own business decisions, i.e., maintain their sovereign rights.* An ability to maintain sovereign rights is logically considering how the tribes have been mistreated and their rights, lands and resources taken away from them (see section 2). This was mentioned by the Harvard Project.
- It is important that no *land is lost as part of developing a business venture.* This was not mentioned by the Harvard Project, but is just as critical for a successful tribal business venture.
- Some *items or resources are not negotiable and will not be traded during a business venture.* This was not explicitly mentioned by the Harvard Project.

It has not been easy for tribes to develop their economies. The U.S. government used treaties to control how tribes develop their economies. The traditional economic activities were until quite recently based only on the resources located on reservation lands. However, the reservation can be a major barrier to travel and developing tribal economies. Many reservations are located in isolated rural areas. These are the only areas that many tribes were able to keep any land.

The Colville Reservation is isolated because of the building of the Grand Coulee Dam. The Colville Reservation is isolated from industrial centers and there is not even a bridge to get across to lands located above the Grand Coulee Dam. There are two ferry boats to handle small amounts of car traffic but they are not suited for large volumes of commercial truck traffic. This makes developing an economy to replace the salmon economy virtually impossible now on a majority of lands on the Colville Reservation. Unemployment remains high on the reservation and future prospects do not look good either.

Ferry County, located on the upper portions of the reservoirs, remains the poorest County in the state of Washington today, with high unemployment and few economic prospects. The region remains one of the poorest in the country, almost 60 years after the Grand Coulee Dam was constructed. Despite some progress in developing business enterprises, unemployment of the Colville Tribe is still high and all socioeconomic indicators show a dismal state of affairs for its people today. Since the 1970's, when federal attempts to abolish the tribe and terminate its entire existence ceased, the tribe has been steadily developing new businesses. Despite the high unemployment found on the reservation, tribes are emerging as important economic players in the U.S. The Confederated Tribes of the Colville was ranked as the top 18th business in the State of Washington in a recent Washington CEO magazine listing. But the tribe still has a long way to go to achieve parity with the PNW U.S. economies.

Tribes developing their business enterprises have also had to contend with mismanagement and robbery of their resources by the government agency set up to provide trustee oversight on tribal resources. The Bush Administration appointed a man named Steven Griles to serve as the Deputy Assistant to the Secretary of Interior. He was the number two federal trustee over Indian resources and their development, only being superseded by the Secretary of Interior and the President. Griles formerly worked for the coal companies and was involved in getting tribes into some bad contacts with the coal companies. He was sent to prison in June of 2007 for unethical and criminal acts related to his duties as Indian trustee in matters related to gaming [101]. Tribal lands are under the purview of the U.S. government who acts in a trustee role. There are many examples of the government benefitting from tribal resources and returning little back to the tribes from whose lands the resources were collected.

The western world recognizes the need to humanize business practices and to more equitably distribute the benefits of economic activities when exploiting resources. The western world is still figuring out what this means and it has been difficult to change a business culture that has formed over several hundred years. There is always considerable resistance to change since the foundation of the western world business model appeared to work so well in the past. But societal demand for including other factors, e.g., societal and environmental factors, when making business decisions, is forcing a radical change in how business decisions are made. The western world dialogue does not mention that they are implementing an Indian business model. Western world business models have a long way to go before they become similar to the American Indian business model. They are on this trajectory even if they don't know that fact yet. How the western world business model is evolving will be briefly discussed next.

## 7.2 Western World Moving towards Humanizing Business Practices

### 7.2.1 Historical Recognition of Need to Humanize Economics

It is widely accepted that business models based only on economic tools are problematic. These discussions are not new and have been heard for many decades. Even historians like Marc Block, wrote "The Historian's Craft" [93]

> "... he never made the mistake of assuming that economic factors explain all human behavior. He knew that man is not entirely rational, that society is held together as much by beliefs and customs as by economic interests."

Many Nobel prizes have been awarded to economists for their innovative breakthrough in how to better predict human behavior in economic models. Despite all this time, writers still write about the need to humanize economics. It seems that we have not successfully achieved this goal and have a long way to go still. Perhaps economic models will never or should not be used to predict human behavior. Therefore an economic model is not a failure if it is unable to sensitively predict human behavior in an economic system. In fact, economics should be only one of the legs of a stool of assessment of economic development potential and success.

Despite worldwide consensus on economics not being the only tool to make resource decisions, it is still the most common and first tool used today. Perhaps we need to change our terminology when we talk about business success so economics is only one leg of our sustainable toolkit.

Economic development is a more appropriate pathway forward and would remove the use of economic assessments as being the primary tool to make business decisions. Economics and economic development require the input of different information to evaluate business success. The excerpt from Schumacher [80] describes the differences between economics versus economic development:

> "Economic development is something much wider and deeper than economics, let alone econometrics. Its roots lie outside the economic sphere, in education, organization, discipline and, beyond that, in political independence and a national consciousness of self-reliance. It cannot be "produced" by skilful grafting operations carried out by foreign technicians or an indigenous élite that has lost contact with the ordinary people. It can succeed only if it is carried forward as a broad, popular "movement of reconstruction" with primary emphasis on the full utilization of the drive, enthusiasm, intelligence, and labour power of everyone. Success cannot be obtained by some form of magic produced by scientists, technicians, or economic planners. It can come only through a process of growth involving the education, organization, and discipline of the whole population. Anything less than this must end in failure."

Currently the western world is working at humanizing economics. It appears to be transitioning economic decisions to include societal welfare. This tool is the Human Development Index [98] and this will be discussed next. It is a tool developed by a consortium of international countries. It is not a tool designed to control how decisions are made at the local or national government policy level. It may influence how some countries conduct business if they want international aid or loans from global organizations. Therefore, it will only impact or trigger a behavioral change in a smaller portion of the global community.

This is the reason why we feel that the Native American business model has to become more generally adopted by businesses. Some companies are already altering their business practices to more closely mimic non-western world models. But these companies are in the minority. Wide-spread and real change will require the "general" public to request and demand this change from businesses. It has to become part of the *western world "culture" and not a fad.*

### 7.2.2 Human Development Index Rankings

Today, western society business leaders want to make better societal and environmental choices. But how do you make better choices in an environment where scarcity is real and equitable business decisions have not been the norm? The recent answer to this problem is to include human development as part of the evaluation process of a countries economic development potential. This evaluation is used to determine which country would receive international aid to develop their economies. The Human Development Index (HDI) [104] ranks countries for how their people are doing in relationship to their economies (see Fig. 7.1).

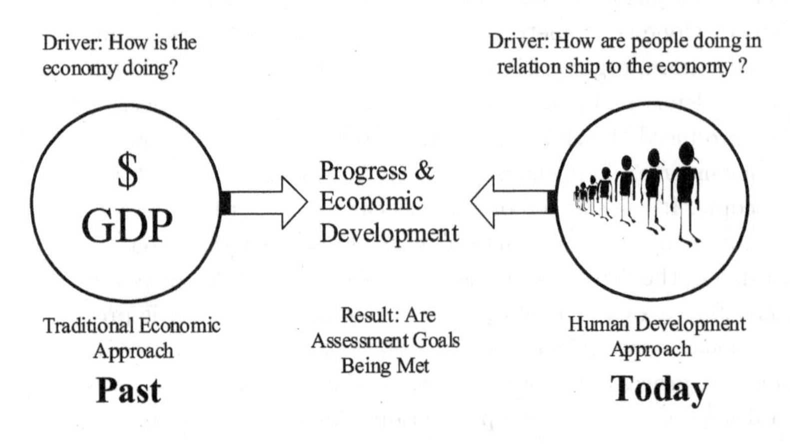


**Fig. 7.1** Schematic representation of the economic approach to measure the economy versus the human development approach.

The following diagram depicts how the economic development metrics have changed to evaluate countries for their development potential (adapted from www.measureofamerica.org/human-development). In the past, the western world metric was mostly an economic approach and asked questions of how well a countries economy was doing. An economic approach does not ask questions such as: How are the people doing? or Are people becoming less resilient or more vulnerable to climate change, environmental or social change as a result of business decisions?

Why is the Human Development Index important to know? If a country has a high HDI it suggests that a countries people are able to adapt to cyclic changes in their environment or society [1]. The key question is whether humans can make sustainable choices no matter what is happening around them? Sustainability in this case means not reducing our choices to adapt to change or disturbances. The idea behind HDI appears to be a country level tool and a way of indexing countries to determine who should receive international funds to develop their economies. A country level index, of course, says nothing about what happens when business leaders make decisions in their regional offices.

Today Europeans attempt to include human development (e.g., education, health, and economics) as part of natural resource business decisions. How commonly it is used in the economic development process and practiced by business leaders is difficult to say. The companies that practice this approach are widely publicized in the media but can be counted on one's hand. The other problem with the HDI is that it is difficult, if not impossible, for business leaders to implement. Most companies cannot afford to focus their business practices to include factors beyond the control of a company. This is why the tribal business model is unique since it is a community decision process overlaid by cultural taboos of what are acceptable practices. Western business models will find it difficult to adjust their practices to include community values. This suggests that ultimately it will be the public that will need to demand these values from businesses. There will be a mechanism by which people can assess whether public expectations are being met. If they are not being met, they can buy other products which would impact the company's business profile.

Even if business leaders want to include human development potential in their business decisions, it would be quite challenging. It is not easy to include humans in any human development index since it needs to include a rationale behavior for how resources are consumed in our economies. It has been difficult to balance business practices with people's behavior since both are complex and very unpredictable. People do not like to follow behavior patterns that follow economic principles even those articulated by world renowned economists. As Krugman [98] wrote,

> "economists fell back in love with the old, idealized vision of an economy in which rational individuals interact in perfect markets, this time gussied up with fancy equations." The key words are "rational individuals" and "fancy equations." Krugman further stated that economists will have to "acknowledge the importance of irrational and often unpredictable behavior, face up to the often idiosyncratic imperfections of markets and accept that an elegant economic 'theory of everything is a long way off.'"

Since the western business model determines what technology is manufactured, it has considerable control over what products are produced by the business community. Funding agencies want to see economic analyses showing that business leaders will recover their investments in a couple of years.

Today, western world economic models determine what products are produced in the renewable energy field. Since business plans suggest it is always more inexpensive and efficient to produce a product if the facility can be made larger and centralized, large-scale facilities are mostly being funded today. However, many large-scale wood biomass facilities have gone out of business because of being unable to maintain a dependable supply of biomass to a mill. A few tribes started to build large co-generation facilities next to a mill but these have not been viable options for them. Since transporting biomass supplies to a facility that is located beyond a 100 km radius from the facility is too expensive, the supply side becomes a barrier. So the western world business model wants to build a large-scale facility while a tribal business model may be more interested in a distributed facility.

Today, tribes cannot afford to take risks to increase their short-term profits. They have adapted to developing markets from a smaller land base while *focusing on economic decisions for the community and not just the individual.* They are better able to adapt to 'boom and bust' cycles of resource abundance and climates (see section 10.5).

### 7.2.3 Beyond Western Business Plans

The large number of different web-based business plans that can be downloaded of the web and used to write a business plan makes it appear to be a simplistic and easy process. It is not. It also propagates the false view that if you get the economic data correct, you will have a successful company. A business plan only provides an economic assessment of a business venture. It does not address all the environmental and social externalities that impact the viability of a business venture.

These social and environmental externalities need to be part of economic development today. Marglin [97] describes how economists assess how to stabilize global carbon levels. He writes

> "...key determinant of the merit of any investment project...is the rate of interest 'charged' on upfront costs incurred to provide future benefits...Much hinges on the choice of interest rate...lower the rate, the more attractive the prospect of preventing global warming...only problem is...analysis misses the point. The important issues are not intertemporal...but inter-regional...likely impact of global warming...different in the rich countries, situated for the most part in the temperate North, than in the tropical and subtropical South..."...
> "...the most salient feature of the global-warming debate is, in the end, how little we know. Probabilistic models shed some light on the future, but quantification breaks down in the face of the overwhelming uncertainty about the effect of the economy on greenhouse-gas emissions on global warming; and, finally to complete the circle, the effect of global warming on economic activity...Given the uncertainty, it makes little sense to rely on a sophisticated calculation of the present value of benefits and costs..."

Taleb [81] addresses the several key rules that businesses need to follow if they are going to be successful under volatile social and environmental conditions. One of his rules is that businesses have to be accountable for the decisions they make. Another rule he suggests is that we should begin to support and favor those businesses that adapt to and learn from their mistakes. The two rules just mentioned by Taleb are provided below [81] because he summarizes well some of our thoughts:

- "At no time in the history of humankind have more positions of power been assigned to people who don't take personal risks. But the idea of incentive in capitalism demands some comparable form of disincentive. In the business world, the solution is simple: Bonuses that go to managers whose firms subsequently fail should be clawed back, and there should be additional financial penalties for those who hide risks under the rug. This has an excellent precedent in the practices of the ancients. The Romans forced engineers to sleep under a bridge once it was completed."
- "Some businesses and political systems respond to stress better than others. The airline industry is set up in such a way as to make travel safer after every plane crash. A tragedy leads to the thorough examination and elimination of the cause of the problem. The same thing happens in the restaurant industry, where the quality of your next meal depends on the failure rate in the business—what kills some makes others stronger. Without the high failure rate in the restaurant business, you would be eating Soviet-style cafeteria food for your next meal out."

These are useful rules to think about when making sustainable decisions since they guide how society may counter the western business model that dominates in the global markets today.

Marglin [97] has several take-home messages. He writes why using economic tools to prevent global warming is difficult because these tools do not address inter-regional impacts and lack knowledge in the areas needed to make economic decisions. These tools also face the challenge that the assumptions being used in the economic analyses appear to be well grounded scientifically when they are in fact not so. In most cases, frequently out of necessity because models can only incorporate a small amount of information, only part of the scientific knowledge is used to produce these analyses. This causes stakeholders to select the science that they include in their assessments. This approach fragments nature and assesses the economics of resources using a very non-ecosystem-based approach. It does not give a manager the information to allow him or her to adapt to disturbances or social and environmental volatility (see section 10.5). A successful CEO does not just rely on business models but augments assessments with other information that influence their decisions. They are systems thinkers and do not rely on fancy equations to make their final decisions [98]. This does not mean that one does not strategize how to mitigate global warming impacts using economic models but this should only be part of the information that is used.

Business decisions continue to be made without considering their social and environmental externalities. For example, global businesses and governments continue to collaborate to control the building of infrastructures and economic enterprises over fresh water supplies. Water is being managed as an economic commodity and increases the disparity between the up-river and down-river people (see section 1.3). The extent to which global societies build large infrastructures to control water supplies is extraordinary. According to the U.S. EPA, about 90% of fresh water flows are altered or impacted by dams, reservoirs, diversions and irrigation withdrawals [99]. In the past governments were able to justify the large water projects they built and operated because of national security issues and their need to maintain viable agricultural industries [100]. It is more difficult for them to justify these large water projects today. But the building of large water projects continues at an unprecedented rate today.

This ability to build large-scale infrastructures, and the technology to manage the distribution of water at any place in the world are stimulating the privatization and globalization of fresh water supplies [100]. As Gleick et al. [100] write in their introduction:

> "Prices have been set for water previously provided for free. Private companies have been invited to take over the management, operation, and sometimes even the ownership of public water systems. Commercial trade in bottled water has boomed. International development agencies that used to work with governments to improve water services are now pushing privatization efforts. Proposals have been floated to transfer fresh water in bulk across international borders and even across oceans...
>
> However, there is little doubt that the headlong rush toward private markets has failed to address some of the most important issues and concerns about water. In particular, water has vital social, cultural, and ecological roles to play that cannot be protected by purely market forces."

When water resources are privatized, water scarcity and its equitable distribution may become a problem. The costs and benefits are not equitably distributed, especially when water becomes a commodity traded in global markets.

Water is tricky to privatize since the source of water is vulnerable to climate change. Lands with abundant water in the past may have scarcity of water supplies today due to climate change. During the Medieval Period, a prolonged drought was linked to the collapse of the Maya civilization [13]. The Maya greatly expanded their economic markets and trade centers when water was plentiful but then collapsed during the prolonged drought. Their societies did not adapt to the changes that occurred in their microclimates.

The lesson that water tells us is that water scarcity is real. Boom and bust cycles are a recurring phenomenon with water and can be found at different times at different locations in the world (see section 10.5). Just that you have abundant water today does not mean that you will have it in the future and be able to meet your demand for it. Water is temporarily owned by a group of people. Even though humans continue to build large infra-structures, i.e., dams, to control and deliver water to those that live many kilometers away from the dam itself, fresh water supplies are not guaranteed and humans do not own it forever.

Tribes do not assess a problem and make decisions where the production, benefits and impacts are disconnected from one another. The western world continues to make decisions where society pays for the social and environmental externalities instead of the producers of the product. The story in section 7.3 summarizes such a story for Iceland.

Tribes accept the fact that identified solutions to problems are not static and should continue to change. According to tribes there is no finite end-point. This contrasts with the western world approach where once a solution to a problem is identified, to change the business plan or to plan corrective actions is difficult.

The story from Iceland in section 7.3 suggests that economics continues to be the primary decision tool used by the industrialized world. This story highlights the trade-offs associated with inexpensive electricity production using hydro-power and jobs. The Iceland story is very recent and suggests that developing energy intensive industries using inexpensive hydro-power was a top priority for the former government. This decision did not consider Icelandic nature and whether the costs to Icelandic environment were worth the benefits from attracting international companies to locate in Iceland. It suggested a single minded focus on economic development without considering social and environmental externalities. It also shows how a government-driven business strategy did not welcome and fought the input of non-economic factors into their economic development strategy. This story also shows how economic solutions continue to dominate in most western world countries. Read this story next.

## 7.3 How the Energy Intensive Business Model Made the Environment and People of Iceland Less Resilient as Told by Raga

Eighty-five percent of the primary energy in Iceland is derived from hydro-power and geothermal resources (/media/orkutolur/frumorkunotkun-1940-2011.xls).

The dependence of Icelanders on fossil fuels is lower than that of other developed countries. Iceland therefore has a reputation of being a model state of environmental friendliness and a good example on how to combat global warming by using green renewable energy sources. Iceland has, furthermore, earned a reputation of having near unlimited sources of green harness able energy, waiting to be exploited by energy intensive industries.

The development of the energy industry in Iceland has been very fast, even though the policy to promote such industry can be traced back to the 1960s. Despite having received international recognition as the green energy island, no environmental issue has received comparable attention in Iceland. The Nature Conservation and environmental communities have focused most of their efforts over the last 50 years or so to constrain Iceland's quest to harness the islands energy resources. They have questioned the size of the energy resources that can be harnessed. It is an interesting paradox how an environmental solution envisioned by the global environmental community is envisioned as the largest environmental threat to Iceland by its local counterpart.

Low costs of green energy from geothermal and hydro-resources, along with low acquisition costs of land, have made Iceland a very attractive place for the aluminum smelter industry as well as other energy intensive industries (e.g., [105, 106], www.invest.is/Investment-Opportunities/Energy-intensive/). Sustainability of energy development in Iceland, and the trade-offs between conservation and ecosystem resilience need to be examined if Iceland's energy production is to be assessed as being sustainable.

Development of the energy-intensive industry (EII) in Iceland began in the latter part of the 1960s. The Parliament formed a committee to negotiate with potential international partners to encourage investments in Iceland using its vast energy resources. The first aluminum company, with a 33-thousand-ton production capacity, and the first large-scale power station (Burfell, southern Iceland) began operations in 1969. The prime objectives of the EII policy were to use Iceland's natural resources to stimulate the development of jobs and economic growth, and to diversify their overreliance on the fishing industry.

Around 1990, Icelandic politicians started a vigorous campaign to promote industrialization for the good of the country and to mitigate climate change. The economical and environmental energy potential of the country was reported to be 30 TWh $yr^{-1}$ for hydro-power and 20 TWh $yr^{-1}$ for geothermal energy by the Ministry of Industry (Althingi [Parliament of Iceland] (1993) Report 1195, www.althingi.is/altext/116/s/1195.html, accessed 8/15/2009).

These estimates have been used widely since that time despite being subjected to increasing criticism by both academia and environmental groups. According to the report only 10% of that energy potential would be consumed by the year 2015, unless energy is exported to Europe or sold to new energy demanding industries operating in Iceland. The Icelandic government repeatedly advertised that Iceland has limitless renewable energy resources, low energy and land prices and plentiful clean groundwater, along with political stability and a low-paid, well-educated labor force (e.g., [105, 106]).

Despite the increasing global concerns related to human impacts on global climate change, Iceland was admired for producing most of its heat and power from geothermal and hydro-power energy. For example, Iceland has been mentioned by many world leaders for their environmental focus—including Al Gore (in April 2008), as well as the editor of Newsweek (Jon Meacham, The Editor's Desk, Newsweek April 14, 2008). In a Newsweek article in May 2008, a reporter named Geir Haarde, the prime minister of Iceland, as the "World's greenest political leader." Alcoa's Jake Siewart told the reporter, "It's almost the ideal place to invest, because of the combination of a highly skilled work force, an open and transparent democracy and the endless supplies of renewable energy."

The EII policy for the last 20 years has been on an economic development focus. Ecosystem and social services are not being delivered by the Icelandic landscape. This focus and the debates will continue because the environmental movement receives little public financial support to focus attention on understanding other factors that need to be understood if the trade-offs are to be sustainable. For example, less than 10 million Icelandic kronur (US$80,000 as of 13 August, 2008) is the combined sum spent by the environmental and Nature Conservancy groups in Iceland. In contrast, the public power companies pay several hundred million Kr to portray their point of view on how 'green' is their energy production. What is viewed by some as green energy, to be harnessed for the preservation of the world's climate and enrichment of people, is viewed by others to be a catastrophic destruction of pristine areas with high ecological and geodiverse conservation value. These differences in views are not likely to be settled soon since the conflicts of the past decade have reduced the trust between stakeholder groups. In fact, direct hostility has increased recently because of the Karahnjukar project and the building of a smelter in eastern Iceland (Gudmundur á Vadi, personal communication; [107]).

The people residing by the river Thjorsa in Southern Iceland have battled for several generations the National Power Company and the government of Iceland for the protection of the river and its treasures. Their quest started in the 1960s when large-scale reservoir was proposed in the highland oasis of Thjorsarver, from where the river itself originates to a large degree. The 45-year long vigilance of the people of the community has fractured the community and decreased the quality of life of its people [108].

Similar story can be told about the people of Lake Myvatn, a wetland area of international importance in Northern Iceland. It is the first designated as a Ramsar site in Iceland. In August 1970 the community of Lake Myvatn managed to halt a large-scale hydro power development in its system when they removed a dam in one of the river outlets of the lake. They used dynamite to destroy the dam.

In the early 1980s a geothermal power station Krafla started operating in the area of the Lake, with the power capacity of 60 MW. A small power plant of 3.2 MW has been operating in Bjarnaflag, ~ 3 km distance from the Lake since 1969. This limited power production has likely already polluted groundwater in fissures between the lake and the power station. These fissures have been popular by locals and tourists for natural hot springs and used for bathing. Recently, the water turned color, from clear to turbid.

A master plan for hydro and geothermal energy resources in Iceland, which is being voted on by the Icelandic government in the winter of 2012–2013, proposes an increase of 265 MW power stations in the area. These would be located at Krafla and Bjarnaflag. The fragile lake ecosystem has been suggested to be in peril by some of the largest NGO's in Iceland. In September 2012, Landvernd, the Icelandic Environmental Association, and Fuglavernd, BirdLife Iceland, called the attention of the Ramsar Secretariat to the development and proposed construction of a new power plant 3 km away from Lake Myvatn. The lake was designated as a Ramsar site in 1977, the first site in Iceland to be listed. The NGO's urged the Ramsar Secretariat to investigate and subsequently communicate on the treats the power plant would put on the lake environment. According to the NGO's , problems relating to pollution released during plant operations are waste water run-off, changes in temperature of groundwater inflow into the lake, and airborne hydrogen sulfide.

During the exponential growth of geothermal and hydro-power stations that occurred from 1997 to 2009, scientists and other members of the public who were critical of the energy policy being pursued in Iceland were repeatedly marginalized or blacklisted by the authorities. This resulted in a situation where scientists became too afraid to voice an opinion on the power policy in Iceland during that decade. The author of this story is one of the scientists blacklisted by the government.

After graduating with a PhD in ecology from Yale University in 2000, I was hired to write an impact assessment on a proposed hydro-power reservoir in the Thjorsarver highland oasis. After submitting my assessment, the legal office of the National Power Company rewrote the assessment so it was quite different from my results and the results of the scientists working on research in the area. Serious negative impacts of the power project were deleted from the report or assessment and were changed to being non-significant. When I questioned the changes that had been made to my report, researchers from both the power companies and members of parliament attempted to blackmail me. I received threats and it appeared that my phone was bugged. There also was an attempt to use the media to generate negative propaganda against me suggesting that I was not a credible scientist. Evidence shows correspondence between the power companies and one of the Ministers where the power companies asked the government to make sure I will not receive a public job in Iceland. This resulted in the State Ombudsman declaring that the government broke the constitution in how it handled my affairs.

The intense face of power development was paused in 2009 as a result of a financial crash in Iceland and a change in government. There is, however, an intense push to begin power development for economic purposes again. Powerful political entities still believe that energy development and building an energy intensive industry will be beneficial to Iceland's economic and socio-economic development. Whether energy development will contribute to Iceland's economic development is not clear. A report from McKinsey & Company [107] reveals that capital productivity in the energy sector is the lowest across all sectors of the Icelandic economy. McKinsey considered having 25%–30% of the capital stock directly or indirectly invested in the energy sector to be detrimental to the overall capital productivity of the Icelandic economy. Despite the power industry providing a foundation for a strong export-based heavy industry sector [109], it has the lowest GVA (gross value added) per unit of capital of all industries in Iceland.

Recently, hydro-power and geothermal resources are increasingly being developed within protected areas, or at least impacting protected areas significantly (e.g., Lake Myvatn area and some areas affected by the Karahnjukar hydro project). New national parks or protected areas tend to be enclosed by areas of potential interest to energy companies (e.g., Vatnajokull National Park) and in some cases (e.g., Reykjanes Peninsula) existing boundaries on protected areas appear to be ignored. Some of the endangered flora and fauna include a high variety of heat-resistant bacteria, potentially of commercial use and still needing to be described [110] and rare plants like the small adder's-tongue (*Ophioglossum azoricum*) found solely on these warm grounds. It seems clear that in almost all cases, areas with high conservation (natural and cultural) value also show good potential for harnessing energy. This situation will make these conflicts continue as choices need to be made. It will continue to cause conflicts between economic development and conservation values held by many Icelanders.

The continuing conflicts between resources and lands continue. We just recounted a story from Iceland but there are many other examples that can be written from around the world. Industrialized countries continue to be interested in the resources found on many native people's lands. These countries are willing to pay a considerable amount of money for these valuable resources. However, not everyone wants to sell bits of their land resources. Native Americans have valuable minerals or resources on their lands but are not willing to give these resources away. Some resources are located in important cultural areas and will not be sold. This means that there will be continuing conflicts over these resources.

## 7.4 Reservation Lands Historically Undesirable but Rich in Economic Resources Today

In the past, resources were taken from tribal lands without tribes having any input of whether this would not happen nor were they compensated for the resources that were taken. Energy was critical for the emerging post World War I economy of the United States. Tribes were often displaced and their lands confiscated when they owned resources that were needed by growing economies. Tribes were forced onto reservations, oftentimes on lands deemed undesirable at the time, but it turns out that many of these reservations contained valuable resources of coal, gas, oil, timber, water, and valuable minerals. Energy continues to be a valuable resource and many Native American tribes are sitting on valuable energy resources that the U.S. and other countries need.

Even today, despite the fact that reservation lands have been greatly diminished by federal government actions, tribes still retain many valuable resources which could become very important to the future of the U.S. economy. For example, Indians own 30% of the strippable low-sulfur coals west of the Mississippi River. Indians own 50%–60% of the uranium resources in the U.S. Indians have 5% of the country's oil and gas reserves [102]. Tribes also own significant stands of timber, with a standing inventory of 0.1 billion cubic meters and an allowable harvest of 1.9 million cubic meters [54].

However, as the abundance of valuable resources on reservation lands became known, this in turn sometimes meant even further land thefts by the order of the U.S. Congress. As an example, the Colville Reservation was created in 1871 but gold was discovered in the same year. The U.S. Congress confiscated these lands known as the North Half of the Colville Reservation. The U.S. Congress compensated the Colville Tribes at a rate of $1 per 0.4 ha for the taken lands. Similarly, gold was also discovered in the Black Hills of South Dakota resulting in the same story being repeated of land theft and a lack of compensation for the gold removed from these lands. The same story was repeated in other parts of the country for various resources, including farm lands and for urban area encroachments. In order to expedite access to these resources, traditional tribal governments were by and large replaced by elected governments with the power to enter into contracts by the Indian Reorganization Act of 1934. Prior to this, many traditional leaders of tribes maintained that resources were not commodities to be bought and sold, it would be akin to selling your mother.

Tribal lands were sought by incoming settlers for homelands and for farms, and also for minerals and energy resources. Pressure was put on the U.S. Congress to open up reservation lands for non-Indian acquisition. Tribes east of the Mississippi River were relocated to the Oklahoma Territory only to be further displaced when the oil corporations of the U.S. wanted access to the rich oil fields.

Similar stories played out in the coal deposits of the Southwest and Great Plains. The Bureau of Indian Affairs did not believe tribes to be capable of developing their own resources. Their solution was to adopt a lease policy, to let resource companies come in and lease Indian lands and tribes would get a royalty for their resources. Whether this policy was well intentioned or not is debatable, but the results are not. There was a wholesale exploitation of tribal resources and tribes saw little benefit. This exploitation still takes place today.

The continuing U.S. government exploitation of tribal resources will be briefly discussed next. This is a continuing practice that still occurs today. It continues the practice of the U.S. government facilitating the exploitation of fossil and natural resources found on tribal lands under the guise of meeting trustee responsibilities and helping tribes to develop their economies. This is changing as tribes begin to control and manage their resources but since this practice continues it has to be mentioned.

## 7.5 Trustee Exploitation of Tribal Resources on Reservations

Some Tribes still have large land holdings, despite genocide and removal policies. Some reservations are rich in natural resources. But most tribes are still mired in deep poverty. The Bureau of Indian Affairs viewed leasing as a solution to this problem. This was their approach to develop tribal economies and resources. Their intentions may have been good, but the outcomes were often tragic. Tribal resources would be leased to outside non-Indian businesses. This solved several problems. The government did not have to put up the capital to invest into tribal resources, and appropriating dollars for Indians has seldom been a national priority. The outside businesses also had expertise and experience in extracting and processing the various natural resources. So, with virtually no investment of federal money, tribal resources could then be developed and this would provide lease payments and royalty payments to tribes who badly needed the income. This was the intent. In practice, things sometimes did not work out so well.

Energy resources were exploited early in history, especially gas and oil fields. According to Parker [103] "The Secretary of Interior and the Bureau of Indian Affairs continually failed to uphold the federal government's trust responsibility in the leasing of oil production of Indian lands during the twentieth century. The BIA frequently did not obtain the highest rates possible, ineffectually regulated leasing, and sometimes leased Indian land without permission... some leases on the reservation lands provided Indians with less than 1 percent of the profits made by the leaseholders in their exploitation of the property." [103]

The Federal government's mismanagement resulted in the losses of millions of dollars of money that was supposed to have gone to the Native American owners. Parker [103] reported that

> "Energy companies underreported royalties due, failed to make payments, or made late payments without penalties being assessed. Theft of oil was reported on the Blackfeet, Wind River, and Navajo Reservations amongst others..."

Energy was an important resource being collected from tribal lands, but tribe received little benefit from these resources. The technology, jobs, and capital equipment remained in the hands of outside corporations [8].

There were several problems. The Bureau of Indian Affairs itself was never adequately funded to carry out its mandated duties; there was always a shortage of administrative capacity to actually serve as a trustee over Indian resources. There was a lack of management and administrative systems. Personnel were in short supply. The system was never set up properly to begin with. The BIA has not been able to keep up with recording land transfers. They also found it difficult to keep track of heirships.

The BIA has lost track of thousands of landowners and millions of dollars of lease moneys. These records have basically been lost and are unaccounted for today. This has resulted in a flood of tribal trust lawsuits versus the U.S. government. Many of these cases are currently still pending. The most famous of which is known as the *Cobell Case*, a class action lawsuit versus the U.S. asking for an accounting of tribal member Individual Indian Money Accounts. This case has been dragging on since 1994 and has led to some contempt of court charges versus several Secretaries of Interiors, who have been unable to settle the case. The plaintiffs contend the U.S. owes billions to Indians, and the U.S. says it only amounts to less than $500 million owed to Indians. There are similar lawsuits in place for tribal assets.

In addition to the incompetent trustee problems, there are other problems related to the leasing policy. Typically the tribes *do not* own any of the plants or facilities and the technology is a black box. So the tribes have no way of knowing whether they are getting fairly paid or not and the BIA typically does not have the capacity to determine it either. Since tribes are locked out, the technology remains a mystery and control stays with the non-Indians. Typically, the tribes are also locked from filling most of the jobs that open up. Few efforts are made to train tribal members to be able to fill these jobs.

Beginning in the 1970s, tribes have gradually begun to assert more control over their own resource development. Tribes have begun to get more organized amongst themselves to share information and to build up a collective expertise to deal with the resource issues. Tribes established several organizations for this purpose. One of the primary resource organizations was the Council of Energy Resource Tribes.

Today a number of tribes have taken over their resource companies completely or else they have much more significant input and control over non-Indian companies if present. Of course the picture is not perfect yet. Tribes often have been left with major environmental problems from decades and decades of BIA mismanagement. Dealing with the cleanup may take many more decades in some cases.

Due to pressure from tribes, Congress passed the Federal Oil and Gas Royalty Management Act in 1982. This legislation included provisions for "new reporting procedures, inspection, enforcement, and penalties... responsibility turned over to the newly created Minerals Management Service...", but thefts have continued to be reported [103].

Despite attempts to correct the situation, problems persist. The trustee continues to struggle and Indian resources continue to be the targets for unethical government and private forces. As recently as 2009, another Indian resource mismanagement scandal erupted in the very agency whose purpose is to protect Native Americans. It is reminiscent of the tales of thievery and exploitation from the 1800s in Oklahoma, apparently some things stay the same.

The following was reported in an industry newsletter from 2008, "A report released by Department of Interior Office of the Inspector General has revealed ethical violations by employees of the Royalty in Kind Program (RIK). The report alleges unbridled unethical conduct by employees, including illicit sexual relations with both RIK employees and members of the gas and oil industry, illegal drug use, and acceptance of numerous gifts and gratuities from oil and gas companies." The report continues, "it was found that 19 RIK employees—one third of its staff—had accepted gifts." The Inspector reported that Chevron had refused to cooperate with the investigation. The Secretary of the Interior Kempthorne stated that he was outraged and promised swift action to fix this problem.

The U.S. government dealing with Indians and oil are fairly well known. Similar tales could be told about other tribes and other resources also. There are numerous books and even movies that touch on this exploitive history. This is the same story that has played out in the Northwest when the demand for energy or water resulted in tribal resources being used or hydroelectric dams, e.g., Colville Tribes and Grand Coulee Dam, being built (see section 2.2.4). In most cases tribal resources were exploited and tribes were not compensated for the full value of the resources taken and used to build the U.S. economy. Tribes lost much when these deals were made. Tribal resources were exploited, their lands confiscated, and their ways of life destroyed. It has taken over sixty years to even get to partial compensation for damages resulting from the Grand Coulee Dam.

### Coyote Essentials

- Economic choices are more than fancy equations or elegant theories.
- The Native American business model forgoes emphasis on maximizing the generation of profits in order to maintain all of their resources.

# Chapter 8

# Portfolio Element IV: Creative Governance from Consensual Flexible Partnerships

## 8.1 Long Western World History: Few Stories of Consensual and Equitable Governance

### 8.1.1 Historical Top Down Governance

The western industrialized countries in general do not have a history of making decisions that are inclusive of the village or community impacted by the choices made. The village used to manage locally-based resources but most of the western world history is a story of a top down approach to natural resource decision-making.

Figure 8.1 shows the long history of rulers, the rich and political elites who made and controlled the decision process in nature. The general public has mostly been excluded from the decision process for natural resources until the

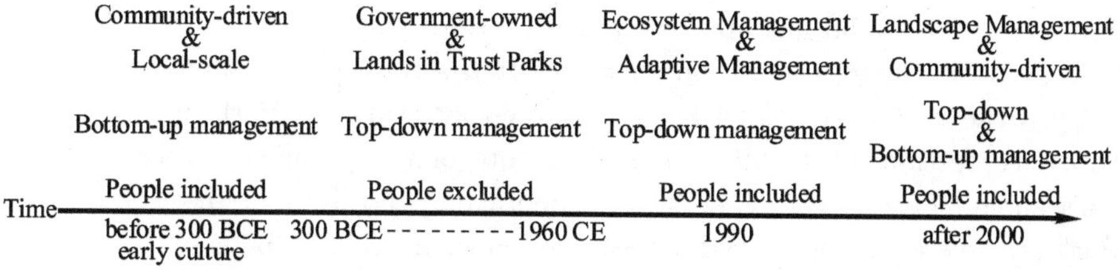

| | before 300 BCE early culture | 300 BCE - - - - - - - - - 1960 CE | 1990 | after 2000 |
|---|---|---|---|---|
| | Community-driven & Local-scale | Government-owned & Lands in Trust Parks | Ecosystem Management & Adaptive Management | Landscape Management & Community-driven |
| | Bottom-up management | Top-down management | Top-down management | Top-down & Bottom-up management |
| Time | People included | People excluded | People included | People included |

Fig. 8.1 A diagrammatical time line for how nature was managed by communities or government.

early 1990s. This means that there is a 2000 year history of top-down management compared to the recent three-decade history of people included in this process.

Top down control of nature and its resources is the most common practice followed in Europe. If the ruler's dictates were not followed, it could result in imprisonment or even death. The most brutal case recorded was in the Old German laws mentioned by the Roman author Tacitus:

> "He noted that the penalty for someone who dared peel the bark off a living tree (and thus killing off the tree) was to have his navel cut out and nailed to the tree and then to be driven round the tree until all his guts were wound about its trunk." [1]

At times a King's decision worked to the future benefits of a country even though that was never the original intent. For example, Louis XIV of France had a Secretary of State for Foreign Affairs, Controller General of Finance and Secretary for Naval Affairs who developed a program of economic reconstruction that made France a dominant sea and economic power in Europe. This was codified under the 1669 Colbert ruling on Waters and Forests (www.foret-de-bourgogne.com/index.cgi?RUBRIQUE=REG). Colbert recognized the need for France to manage its waters and forests to become a ship building and naval power while maintaining the lavish life style of the King, Louis XIV.

When the 1669 Colbert ruling on Waters and Forests was first implemented, it was not a democratic decision process. It was a King's decision to move forward on this rule. This rule eventually did benefit France. It stimulated the planting of vast oak forests that has allowed France to provide a valuable commodity on global markets today, e.g., oak barrels that are preferred by wine makers around the world for making quality wine.

When the first European colonialists arrived in North America, their decision-making systems followed the European model of Kings or rulers who were the ultimate decision-makers. Despite European colonialists and their descendants adopting a federalism model of governance in North America, a process for making decisions on more conflicted and competitive resource issues did not exist then. There did not appear to be a need for such a decision process since scarcity of resources was not on anyone's radar screen at that time.

Even the first national park, i.e., Yellowstone National Park formed in the U.S. (see section 2.2.1), did not seek a vote of the public to determine if a park should be formed. The Native Americans who lived in the Yellowstone Park area were also not asked if they wanted their lands to be converted into a park.

---

1 //www.smr.herefordshire.gov.uk/education/Medieval_Countryside.htm.

### 8.1.2 Historical Western World Governance Structures that Did Include People

For the western industrialized countries, the northern European countries (e.g., Scandinavian countries including Iceland) historically practiced a governance structure similar to Native Americans. Iceland's historical consensual practices based on community decision-making are summarized in Box 19.

**Box 19: Community Decision Process in 9th to 11th Century Iceland**

Since the late 9th century and until 1262, Iceland was an independent country with no centralized governing body. Legislative and legal affairs were settled in assemblies, which were both regional and in a centralized manner. The centralized assembly, named Althingi, was held in mid to late June for 2 weeks every year. This assembly was first held in the year 93 CE. In the center of Althingi were 48 members, named godar. Each godar selected two additional parilamentary members. In addition, there was the logsogumadur, or lawspeaker, elected for a term of 3 years, who was the only paid member of the assembly, and who had the greatest authority. The lawspeakers job was to Chair the meetings and proceedings, and know by heart the laws of the country. The authority of the lawspeaker lasted only for the duration of the Althingi. Rulings in matters that needed judical attention were grouped into three main catergories: 1) payment of penalties, 2) banishment for 3 years and 3) banishment for life. There was no death penalty per se, but those banished for life were considered rightless in society. The assemblies had only a legislative and jurisitcal domsvald power, but no administrative power or framkvæmdavald. In fact framkvæmdavald was in the hands of the relatives of the victim, creating a sort of vendetta system, which could be enforced on the perpetrator or someone closely related to him.

The laws of the lawspeaker were written down in 1117–1118 in Iceland. This created a law book, Gragas, which is one of the most ancient texts of written Germanic law. For some reason the law book was given the name of one of the two most common goose species in Iceland, the grey lag goose (*Anser anser*). The book is a very good reference to the way people ruled the country during its earliest centuries, and describes the way of life in Iceland at that time. It is also a very good reference for how decisions were made on land management and use of resources.

In the land and natural resource use chapter of Gragas, detailed rules are laid out for how grazing lands should be managed, and how forests should be harvested in a sustainable manner. These rules were especially important

(Continued)

in cases where the resources were owned by more than one owner. Common lands should only be used to graze a certain number of sheep, and the grazing land in the commons should not be cut for hay. Also the rules apply in cases if the environmental status of the commons is degraded, or someone suspects that someone has fatter sheep despite grazing at a lower intensity. Changes in land use regimes where brought to the regional assemblies for decisions. Grazing of pigs were forbidden on common lands. Hindering the migration of fish by putting nets across rivers was forbidden, and subject to penalties.

In later centuries, Iceland developed new law books. These books were written when Iceland formed formal alliances, first with Norway and later with Denmark, and after a series of famines and plagues. The former sustainable systems that were part of the earliest laws deteriorated were not retained when these new laws were written. This change was reported to result in the degradation of both forest and grazing land resources.

Despite this historical example of a consensual governance structure practiced in Iceland, most western European countries did not follow a practice similar to what occurred during the early history of Iceland. It is also noteworthy that despite Iceland practicing this form of governance for about 400 years, they did not continue these practices after 1262 (see Box 19).

It is not until much later, beginning in the 1980s, the western industrialized world recognized the need to link the social and the natural systems when making nature decisions. Therefore, humanizing natural resource management decisions is a very recent phenomenon in the western world. Ecosystem management became the term to use when making decisions for our environment and natural resources [111]. This new paradigm highlights the importance of including a broader sector of society in making decisions. The tools to practice this coupled natural and social systems, however, did not exist when ecosystem management was initially adopted by U.S. resource management agencies in 1992. The western world was a late entrant into formally including humans as part of the decision-process for their natural landscapes. They still are struggling with developing the tools needed to implement this emerging concept.

The western world diagrammatically represents the ecosystem management concept as a circle representing the social system intersected by the natural system circle where the circles overlapped [112]. This approach suggests that the social and natural systems only need to be linked in the portions of the circle segments that intersect. It also suggests parts of the ecosystem are not connected to the social and natural systems.

This circle approach made it easy to identify social and natural science factors or indicators that could be included in a list when making complex environmental decisions. This approach is less able to identify the causal drivers or links between the social system and natural system. The concept of interconnected land systems identifies generic drivers of change but is difficult to use this model at any location. It is less able to connect society to changes occurring in the ecological system. This means that it is less able to detect when the social system may become vulnerable to decisions made in the environment.

The social system is also more complex than the figure would suggest. This conceptual representation of intersecting social and natural systems does provide a framework for thinking about a resource problem. It does not lead one to know how to practice real ecosystem management. It is difficult to include all the causal factors that one may need to know when practicing ecosystem management. It would be a decidedly messy diagram and not very helpful to a practitioner.

In an effort to develop consensus among disparate stakeholder groups, many attempts have been made to forge a collaborative group of all the relevant stakeholders. Typically, all stakeholders are brought together into the same common space to work through the conflicts. Even though this has been a successful model of collaboration, there are few examples of long-term group collaborations that survive beyond 10 years. This highlights the difficulty of bringing different stakeholder groups together who all have a vested interest in the results and want different solutions.

When resource scarcity and the competitive values from the same natural landscapes became a forsworn conclusion, the western world recognized the need to collaborate on decisions made for natural resources and nature. The approach taken by the western European and their descendants was to form internationally-based organizations with representatives from each member country to make decisions for the global society.

Many of the initiatives developed were started by non-governmental organizations. These organizations represented many countries and advised governments on international law and policy formulation on multiple topics [113]. The United Nations was formed in 1945 and has 193 member states. Many of these organizations focused on developing cooperation among member countries on issues that needed to be addressed by global citizens: peace, conservation and biodiversity, economic and social development, human rights, etc.

Most of the global multilateral organizations were formed after World War Ⅱ. They contribute today to global problem solving because they [114]

> "have become unique repositories of specialized knowledge gathered through decades-long worldwide operations. But their roles and their relationships to their owners and overseers... preclude them from taking on a central role in global problem solving. They cannot just

> decide on their own to cut through the tensions and disagreements among their owners."

Rischard wrote a book entitled *High Noon. 20 Global Problems 20 Years to Solve Them* in 2002 that introduces new thinking that is needed to solve global problems [114]. In his book, he writes how urgent global problems are getting worse and global communities have not developed effective or adequate strategies to deal with them. He recommended "networked governance" since "undeniable *failure* of the entire international setup and the world's nation-states at the task of fast and effective global problem-solving" [114]. He clearly states that this cannot be a process where we "reform a few existing institutions. One of the tragedies of our times is the widespread belief that we need only reform a couple of existing international organizations, and presto! All will be fine."

We would take Rischard's [114] ideas even further if the global community is to shift natural resource consumption decisions from being controlled by special interests groups. We propose that the "federalism" idea introduced by the Iroquois League has merit and would allow a "village" or the "local placed" context to be part of the decision-making process. Rischard [114] articulated well the importance of knowledge and creativity to solve global problems today:

> "Simply put, earlier technological revolutions had to do with transforming energy or transforming materials. This one has to do with the transformation of time and distance, and thus cuts deeply into the fabric of society. At least as important, it has made knowledge and creativity the number one factor of production—far more important than capital, labor, and raw materials."

As summarized on the book flap of Rischard's 2002 book [114]:

> "Rischard proposes new vehicles for global problem-solving that would be acknowledged by governments but that would function as extra-governmental bodies devoted to particular problems. Their powers would not be legal but normative: They would produce globally recognized standards and would single out the nations and organizations. . . "

Books have been written on the topic on how to bring polarized and conflicted groups to the same table to work on particularly tricky environmental and resource problems. Several years back one of us (Kristiina A Vogt) was at a consensus building workshop where several critical insights became clear while playing out consensus development scenarios. It became clear that if a conflict continues unabated, the really important topics had not been 'put on the table' for discussion. Another insight that emerged during this workshop was that the negotiated solution will not succeed when a rival group feels that they have to give up too much for the benefits they derive from making the agreement.

This recently happened in Klamath Falls, Oregon where a collaborative group worked on a Klamath Basin Restoration agreement in 2010. When this agreement was reached, it was reported [115]

> "after five years of confidential negotiation, an unlikely alliance of Native American tribes, environmentalists, farmers, fishermen, governors and the federal government signed the Klamath Basin Restoration Agreement. The agreement was hailed as evidence of a new era in the West in which bitter divisions over natural resources could be bridged. Within a decade, it dictated, four dams would come down, enabling much of the river to flow freely and its once-mighty run of salmon to return. At the same time, farmers would be assured of water for their crops and affordable power. And Indian tribes would regain land lost decades ago."

The Klamath Basin Restoration Agreement appears to be unraveling two years after it was negotiated. Some farmers became afraid that they were losing access to sufficient supplies of water. They pursued political solutions to control their access to this water. Since the dams collect much of the water used by farmers, dam removal made the farmers worried that it would reduce the amount of water that they would be able to access. Since about 74% of the surface water in the western U.S. is used in agriculture (irrigation and livestock) [116], this is a valid concern. It became apparent that that dam removal was insufficiently addressed during the negotiations. Or, the farmers did not articulate their concerns sufficiently when the Agreement was originally made.

It also could be that this competition for water supplies has been aggravated by the droughts that this region has experienced. Droughts have "prompted federal water managers to shut off irrigation to ensure enough water for endangered fish." [113] This has been further complicated by the "commercial salmon fishing on the West Coast was shut down in part because of the decline of salmon populations in the Klamath. Scientific research indicated that removing the dams was the best way to save the salmon"[115]. The tribes also have treaty rights to fishing and hunting from these lands that they had relinquished to the federal government as part of treaties. The article written by Yardley (2012) wrote [115]:

> "I always refer to us as the radical middle because there's nothing radical in the Klamath about fighting over water," said Craig Tucker, the Klamath coordinator for the Karuk Tribe of Northern California and a supporter of the 2010 agree. "What's radical is learning how to share."

The European colonialists had much to learn from the American Indians they came in contact with how to communally share resources. But their minds were not receptive since they felt they had manifest destiny on their side (see section 2.2). There was no need to consider consensual sharing or collaborating on resources that were being exploited. It is no accident that the Grand Coulee Dam dispossessed the Colville Tribes of their lands, isolated them from the rest of the world, subjected them to poverty, and that the benefits of this project flowed to incoming non-Indian farms and businesses. This is consistent with the colonialistic policies which have been in place since 1492 when Columbus landed in North America. Laws, policies, and history itself were not implemented for the benefit of indigenous peoples [117]. The intent was to exploit tribal resources and dispossess tribes of their lands and resources as expeditiously as possible. One observer, remarking on similar circumstances in the Pine Ridge Reservation remarked in the 1950's, that the BIA was, "The Company in a company town."

"The bureau dominated the economy as employer, purchaser, and consumer. It handed down the laws and ran the police and courts. It controlled the tribes, only economic asset... they controlled everything. You couldn't even sell your cows without permission..." [117]. Instead of protecting tribal assets or at least getting a fair deal out of the Grand Coulee Dam project, the BIA had a practice of assisting with the dispossession of tribal lands and would tell Indians that you cannot stop progress and if you don't sign on you may end up with nothing at all. But, "... history is about power. It is mostly about power." [72]

The Grand Coulee Dam for example is a story of man conquering nature, a wonder of the world. There are many happy farmers with water to irrigate their farms and tourists with big lakes to water ski on, and happy housewives with modern electric appliances. The Tribe is totally invisible and there is no mention of the lost salmon or beautiful Kettle Falls, these parts of history had vanished. It took the Colville Tribe half a century to combat this history and to get a settlement against the U.S. government. The largest settlement of its type was negotiated and approved by Congress in 1994 and is known as the 181-D settlement. It compensates the Colville Tribe for lands flooded based on a formula derived from annual power sales. There was no compensation for fisheries loss, culture loss, economic losses, or any other losses, however. The U.S. government initially offered the Tribe $63,000 in 1946. The Tribe immediately filed a claim in the U.S. Court of Claims, which was set up by the U.S. Congress to deal with these issues [118]. The court expired in 1978 and the Colville case was not settled. The Tribe had to go directly to the U.S. Congress and reopen the case later, when it had the financial resources to do so.

Hydro-power is often cited as the perfect power source, it is both clean and renewable. However, even hydropower can create winners and losers, some benefit and some pay the price. Interestingly, the 1990's is cited as a time when indigenous scholars started to assert indigenous views and created indigenous history movements [119] and Native Americans began to assert their power. Tribes began to articulate their own stories. The Colville Tribe produced their own documentary and story. They went to DC and presented their own story to the U.S. Congress. They bypassed the designated trustee, the BIA.

There are still many vestiges of the colonialistic policies within the federal government. These have been developing since the founding of the U.S. and they need to change. Many would argue that this policy was a pendulum and sometimes swayed with sympathy toward trying to assist Tribes, but the end result cannot be denied.

Tribes have lost the vast majority of their lands and Tribes, with some exceptions, are still deeply mired in poverty. The U.S. policies set up to expedite this exploitation are still in place and in many cases need to be changed if Tribes are to progress and maximize the benefits from their own resources.

## 8.2 American Indians: Village and Confederacies Make Natural Resource Decisions

Individual tribes have many different governance structures so the structure by itself does not produce or result in equitable natural resource decisions. Tribes have a history of equitable resource decision-making despite having a multitude of tribal governance structures. This supports our contention in this book that culture and traditions are most important for setting the context for sustainable and equitable decisions. You do not need to adopt one model for how to build your political structure but you need to adopt a decision-process that is consensual and where multiple people provide leadership on how any resource is produced or consumed (see section 10).

Tribal governance structures are highly variable and reflect past and current approaches to decision-making bodies. In 1934, the Indian Reorganization Act was passed by Congress which had as one of its goals to eliminate traditional governance structures of American tribes. There was a movement to get rid of traditional Chiefs who had life-long appointments in many tribes and to establish an elected Council or Boards type government. But a few tribes retained their traditional government forms. The Pueblos in the Southwest U.S. kept their traditional systems.

Some tribes adopted hybrid governance structures. The Warm Springs Tribe has an elected Council, but they also have three Chiefs who hold lifetime appointments. This contrasts with the Colvilles where a Chief has a two-year term.

Other tribes have changed their Constitutions to lengthen the terms of their leaders or chiefs, so some changes are slowly taking place. Therefore it is difficult to make easy correlations between forms of government and success in how resources are produced or consumed.

Tribes also have very different resource bases on reservation lands that can be used to develop local economies. Further, some of the traditional based governments do not place the same value on money success.

There is disagreement in the literature on the full extent that Native Americans contributed to the development and writing of the Constitution of the U.S. [120]. Some historians suggest that when the first European settlers came to eastern North America and formed 13 separate states, the Iroquois Chief Canassatego suggested the formation of a federal state comprising these same 13 states and what policies would allow this to occur [121]. Benjamin Franklin was one of the advocates for the adoption of the policies embedded in the Iroquois League. "Unlike European governments, the league blended the sovereignty of several nations into one government... The Indians invented it (federalism) even though the United States patented it." [121].

We really do not think it is as important to know how much of the U.S. Constitution was derived from Native American governance policies. We will never know this information and can only speculate on this topic. Historians have not developed any consensus on it over the last several hundred years. We need to move beyond these arguments and see what aspects of Native American policy are worth adopting. It is more important to discuss how the American Indian practices of governance can inform us to humanize resource use decisions today.

So this takes us back to looking at how decisions are made and not the governance structure itself. Tribes have been especially effective at avoiding polarized situations because the village is part of the decision process. They do not practice a "top down" approach to making decisions. So what elements of the Iroquois League are needed to expand the "networked governance" and "global issues networks" proposed by Rischard [114]? We suggest that we need to first shift towards a "bottom up" approach to decision making on scarce resources, i.e., consensual governance and not governance driven by special interest groups.

We suggest that American tribes have much to contribute towards learning how to practice consensual governance. The designation of salmon chiefs by tribes to arbitrate conflicts over salmon suggests an approach that should be pursued by today's natural resource managers. It was the responsibility of the Salmon chief to make sure that someone with no food will get some salmon. Chief Thompson was the Colville salmon chief who presided over Kettle Falls and was an arbitrator when conflicts arose over salmon.

The following excerpt describes Chief Tommy Thomson well and the role that he played in the equitable distribution of salmon to ensure the viability of salmon into the future (See Fig. 8.2):

> "When I knew him he said he was a hundred years old and he probably was. He was quite dignified but a very tiny man and very quiet. His spokesman was Flora and Flora wanted everybody to treat him with respect. She looked after him very carefully. Before that and even to that time—which I think is remarkable looking back—he was the one who designated the rocks where you would fish. Now, some were better than others but that rock belonged to you and you [fished] that one but you didn't [fish] anyone else's. And he controlled the Indians to keep that and they looked to him as the chief there even though they came in from other tribes. They had great respect for him in many ways."
>
> ~ Barbara MacKenzie, interviewed by Katy Barber,
> 30 September 1999 ~

**Fig. 8.2** Chief Tommy Thompson at Celilo Falls, Oregon.
Item No: Williams G: NA_Thompson, Courtesy Oregon State University Archives.

These chiefs mentioned others in the tribe to carry on their traditions. Equitable resource distribution was not a responsibility that ended with the death of a Chief. The Chief took steps to ensure that resource responsibilities continued beyond his or her leadership. This continuing mentorship of future leaders, who are targeted for their future leadership potential, is described in Box 20.

**Box 20: Mike's Encounter with Eightball Jim**

One of the iconic tribal leaders in the Pacific Northwest was Nathan "Eightball" Jim, from the Warm Springs Reservation in the state of Oregon. I was a Councilman from the Colville Confederated Tribes in Washington State and had arrived early for a meeting with the Columbia River Intertribal Fisheries Commission at their Portland Office. Eightball was one of my mentors, he was looked up to by all tribal leaders, and he was happy to see me, he grabbed my arm and said, "Mike come with me, I need to talk to you." We sat down in one of the offices and he proceeded to tell me a story. He said, "Mike, when I was a small boy, I was the young protégé for the Celilo Falls Salmon Chief." Celilo was one of the best salmon fisheries in the world for thousands of years and was located on the lower Columbia River. It supplied the salmon which provided food for thousands of Native Americans. The Salmon Chief was in charge of everything at the Celilo Falls.

Eightball said that in those days the Indian people knew the salmon very well and by looking at the salmon they could tell where it was heading to. He went on to say that when the Salmon Chief spotted incoming salmon that were heading up river to the Kettle Falls Fishery, that he would call an immediate halt to all fishing at Celilo Falls. This was to allow passage of the traveling salmon so that they could make it up to the headwaters of the Columbia River to ensure that the upriver tribes had salmon to eat and also to ensure that enough salmon made it up river to spawn and thus sustain the great salmon runs forever. Likewise, the great up river Kettle Falls Fishery also had its own Salmon Chief. He was in charge of all that happened at that fishery. At the right times, the fishing would be stopped to allow fish to pass though to the headwaters of the Columbia River. These two great salmon fisheries were managed and sustained for thousands of years over distances of hundreds of miles and amongst tribes who were often at war with each other. They put aside their differences to manage and sustain the great salmon runs.

Eightball was dead serious in telling this story, he was not laughing and joking as he often did in other settings. He said he has spent his life trying his best to protect the river. To undo the damage from the dams. To protect and bring back the salmon. To fight for our sacred right to fish as had our ancestors for thousands of years. He said he was getting old and soon it would be up to the new generations to continue this work. This is why the creator put us on this earth.

(Confederated Tribes of the Colville)

## 8.3 Link Taboos to Non-negotiable Values When Making Economic Decisions

Today, many people benefit from indigenous peoples who valued and held water and their lands sacred (see section 1.3). This continues to support our metaphor where up-river people established sacred sites hundreds of years ago to protect water. Today's down-river people are the beneficiaries of these past practices. Today we need to change this disconnect between where resources are produced in the headwaters of rivers and down-river political powers who consume these resources in the lowlands. The potential of reconnecting resource production and consumption as being positive for sustainability is shown in the following fact —about a third of the fresh water drunk in large cities in developing countries originates from these sacred areas established by villages a long time ago [23].

Sacred groves with ecological significance were established by past societies and they conserved multiple aspects of nature [122]:

- Conservation of Biodiversity—The sacred groves are important repositories of floral and faunal diversity that have been conserved by local communities in a sustainable manner. They are often the last refuge of endemic species in the geographical region.
- Recharge of aquifers—The groves are often associated with ponds, streams or springs, which help meet the water requirements of the local people. The vegetative cover also helps in recharging the aquifers.
- Soil conservation—The vegetation cover of the sacred groves improves the soil stability of the area and also prevents soil erosion.

It is how these villages were able to protect their resources that are interesting for our story. They established protected areas where resources production and access were locally controlled by villages, i.e., small, flexible societies, and where the village established rules to protect these sites (see section 1). The villages did not separate their sacred sites from all of the natural resources they controlled. Sacred sites never became a "zoo" (see section 5.2) where humans could look at but not touch nature. They were effective at maintaining the village values in these protected areas by community-accepted cultural taboos and taboos. These taboos were passed down from one generation to the next using folklore or oral history.

The story in Box 21 introduces the idea that each village has to contribute something towards the common goals, i.e., to have salmon in their streams and rivers. In this story, each village had to give something important and valuable, i.e., a beautiful woman as a wife for coyote. This shows a consensual decision process for determining who would be able to fish from the river (see Box 21). There was a need to contribute something to someone else to ensure access to an important cultural resource.

**Box 21: Coyote Breaks the Dam and Brings Salmon up the Columbia River summarized by Mike**

In ancient times monsters had dammed up the mouth of the Columbia River where it enters the ocean. The people upriver were starving. There were no salmon. The main food of the people was the salmon but there were no salmon. So the people prayed. This is one of the creation stories of the upper Columbia River tribes, known as Coyote Stories. Coyote the trickster is the progenitor, sometimes he does good and sometimes he does bad, and as often as not he acts foolishly. Coyote went down to remove the dam to save the people. Using his trickery he was successful. With his staff in hand he led the salmon back up the Columbia River. He made his way up the Okanogan River. He reached the branch of the river known as the Similkameen River. He asked the people there for a wife. They refused. The women there laughed at Coyote. In vengeance, Coyote built a great waterfalls there to stop the salmon forever from going up the Similkameen River. Then he went to Lake Osoyoos. There the people were generous, they gave him a wife. To this day, the Lake is rich in salmon. He got tired of this wife and went up to Okanogan Lake. Again they laughed at him. They got no salmon either. Coyote then went to the Sanpoil River. They wanted salmon. They gave him a beautiful young maiden for his wife. He rewarded them with the finest salmon. But he got tired of this wife too. Then he went up the Columbia to the where the Kettle River enters the Columbia. These people were kind and generous. They gave him the most beautiful maiden for his wife. Coyote built a great waterfall there, Kettle Falls, and built a great village. Coyote said all the people will have to come and beg salmon from them in the future. They will bring many kinds of food to trade for your salmon and your people will be wealthy and have great power. Then Coyote got bored. Coyote left his salmon there and went away. The people's prayers had been answered, if they were kind and generous to Coyote. But if they were greedy and did not want to give Coyote a wife, then they got no salmon. Each river was affected differently by Coyote as he made his travels. This is how things were for thousands and thousands of years after Coyote led the salmon up the Columbia River. The River was the source of their power and food. So long as they prayed and took care of the River and followed their traditions life would be good. Each spring the People pray for the salmon to return and they conduct a First Salmon Ceremony to pay their respects to Mother Earth and the Creator Quillenchooten.

(Confederated Tribes of the Colville)

The key element of sacred sites was that it was locally based and not a bureaucratic centralized policy body that made decisions and was disconnected from its local context. Similarly, the Native American tribes do not develop super centralized governance infrastructures that are harder to mobilize in response to disturbances or altercations. Large centralized governance structures tend to develop specialized decision bodies, e.g., a forest service, health, education and agriculture departments or agencies to name a few. These specialized organizations are mandated to reach goals for each resource under their jurisdiction. This lends to propagate the up-river and down-river approach to decision-making. The communication networks among these different organizations are also difficult to maintain. Frequently decisions are made in isolation from the other organizations or agencies that should be involved in making decisions. All partners do not participate in because of jurisdictional boundaries that stop this collaboration. This is one of the points that Rischard [114] was attempting to resolve when he wrote his book. Many large organizations have a strong culture that is resistant to change even when facing human or natural disturbances.

The idea of having many smaller groups of organized and communicating with decision-makers is not new. In 1973, Schumacher wrote about it in his book 'Small is Beautiful: Economics as if People Mattered' [80]. The ideas in his book have not caught traction in society despite the fact that he made very strong arguments for not maintaining a large centralized economic structure. These issues continue to be debated today, especially in the bio-energy sector. However, large centralized energy biomass facilities continue to be advocated by U.S. federal agencies and their funding agencies. They still consider this option to be the most economical despite the case that can be made for decentralized energy production in rural areas [123].

The world is globally and totally interconnected in a myriad of ways that most of us are not aware of. This would suggest that decentralization decision-making institutional structures will not work to address rural problems. We contend that decentralization can be more sensitive and responsive to an unpredictable climate and resource abundance.

There is a role for the centralized structures but they will not understand the local context sufficiently to most efficiently address regional problems. The decentralized structures should be more adaptive and responsive to local needs especially under conditions of volatility or "boom and bust" cycles. Both scales need to partner on addressing local natural resource issues especially under a climate change umbrella.

The conflicts on climate change and its impacts on vulnerable communities around the globe reflect this challenging issue. Traditional knowledge has to adapt to events that occurred in a different century and/or space. The Classic Maya civilization and governance were unable or unwilling to adapt to climate change [124]. When they experienced an unprecedented high rainfall, they expanded their political centers. This came crashing down on them when this period of unprecedented high rainfall was followed by a almost 400 year period of drought [124]. As Kennett et al. summarized [124]

> "...a drying trend between 660 and 1000 C.E. that triggered the balkanization of polities, increased warfare, and the asynchronous disintegration of polities, followed by population collapse in the context of an extended drought between 1020 and 1100 CE."

In contrast to Native Americans, the Maya Civilization did not have a tradition of learning from the past or maintaining a strong oral history that persisted over the cycles of good and bad climatic periods. They were not adaptive and did not modify their cultural practices over long time periods, i.e., making decisions for the 7th generation.

Today, Tribes link science developed by non-tribal peoples to their cultural practices as part of their adaptation to a new world. The western Tribes are actively addressing climate change and how it will make them more vulnerable to its impacts. One Washington State coastal tribe is already being impacted by rising sea levels and storms that forced them to request alternative lands for their reservation from the U.S. federal government. They needed to move their reservation to higher ground or their villages would be destroyed following a major storm. Many PNW U.S. tribes have climate change plans and are actively planning for the future. They are very aware of the fact that their tribal cultural resource like salmon is especially susceptible to changes in the climate. Tribes are mitigating and restoring habitat to improve the survival of cultural foods especially threatened by changes in land-uses and climates.

Tribes have traditional knowledge and culture that are integral to decision-making that impacts individual tribal members (e.g., where to put the septic system) and the tribe (e.g., what will be the dominant economic activity in the tribe and how does it impact cultural resources?). Tribes have what western world scientists call a bottom up approach in ecology compared to the top down approach used by the western world. For tribes, decisions are not made in isolation but include all tribal members who vote on important decisions (e.g., water, food, energy, forests). The western world has compartmentalized decision-making in organizations specifically set up to assess, monitor and make decisions for disciplinary based problems (e.g., FAO, WHO, UN, etc). Tribes have an easier time to form partnerships with a broad group of diverse people since decisions are not compartmentalized like what happens in the western world model.

Tribes have a multi-level environmental governance structure and do not have one organization making decisions. It is the village or community that makes decisions. The western world models produce 'special interest' groups that dominate the decision process and take advantage of lack of information to confuse the public on complex issues. Special interest groups are more difficult to form in tribes since a village or community making decisions makes this process more transparent, i.e., it would be hard for a special interest group to dominate.

Unlike the long history of excluding people in the decision process for natural resource uses, Native Americans continue their community-driven approaches to decision-making on natural resources. This occurs despite tribes being embedded in a landscape that continues to exclude the "village" and to support a more robust decision process. According to Weatherford [121]

> "This Indian penchant for respectful individualism and equality seems as strong today in Fargo, North Dakota, as when the first explorers wrote about it five centuries ago. Much to the dismay of contemporary bureaucrats and to the shock of the Old World observers, Indian societies operated without strong positions of leadership and coercive political institutions."

Some of the first European chronicles of the Americas support the more equali- tarian decision processes followed by Native Americans. Weatherford [121] wrote:

> "For the first time the French and the British became aware of the possibility of living in social harmony and prosperity without the rule of a king."
>
> In his essay [written in the 16th century] "On Cannibals", Montaigne wrote they are "still governed by natural law... the so-called savages lived better than civilized Europeans... Technologically simple Indians usually lived in more just, equitable, and egalitarian social conditions."

Native Americans, despite being confined to reservations, are active in regional planning efforts for the environment. Regionally, they are implementing and collaborating with non-Indians on salmon and wildlife restoration efforts. Tribes are restoring habitat, redesigning landscapes and even advocating for the removal of dams to make their customary lands capable of supporting their cultural resources. These efforts would probably not be happening without the tribes' ability to push their agenda forward because of their political status/treaty rights.

These stories are briefly described in the next part of our book. We cannot do justice to the many projects that tribes are implementing. But we want to introduce a few projects so the reader will be aware of them and perhaps the reader will be encouraged to do some of their own research on this topic.

**Coyote Essentials**

- Getting the administrative governance structure right is not as important as getting its functions right.
- The governance structure has to be able to adapt to climate change or societal collapse is possible.

# Culture as the Core of Native American Resource Leadership

## Chapter 9

# Traditions Are Not Just Writings Found in Library Archives: Native Americans Driving and Controlling Resources Today

Tribes are actively implementing and allocating funds to projects to restore the major riverscapes in the Pacific Northwest U.S. They want to restore some of their cultural traditions and way of life that they lost upon the arrival of the European settlers to North America. They know that they will not be able to go back to what is described in the next paragraph but are dedicated to restoring as much of their cultural traditions as possible.

> The Columbia River in the U.S. Pacific Northwest has its headwaters in British Columbia, Canada, and travels southwest to the Pacific Ocean. Today it is a series of major hydroelectric dams that provide electricity for the region. But, prior to the dams, the River was one of the world's richest salmon fisheries that supported thousands of Native Americans for thousands of years. There were two great fisheries on the River. The upper fishery was the once great Kettle Falls. It was inundated in 1942 with the completion of the Grand Coulee Dam. The once great lower fishery was Celilo Falls. Kettle Falls was famous for its basket traps and salmon spears. Baskets up to 30 feet long were placed at the foot of the Falls. Thousands of salmon trying to leap over the Falls would fall into the big baskets. Early observers with the David Thompson Expedition counted 2,000 salmon coming out of a single basket trap in one day. The spearing was done from scaffolds jutting out from the Falls. The Celilo Falls fishing is noted for its big dipnets, which are still used today. Actually all forms of fishing technology were used, including weirs, traps, nets, and hooks, but these were the main methods. Fishing itself was done by men. There were strict traditions and ceremonies that had to be followed to ensure that the salmon runs would come back each year. Salmon fishing villages would swell in size each salmon season. Thousands of Native Americans would gather

> each salmon season and included traders and craftsman and visitors came from all over. Salmon supported the local tribes but they were so abundant that they were also a trade commodity. Salmon was dried and put into bales and tons of salmon would be stacked all around each fishing site. This salmon was brought by canoe downriver and traded up and down the Pacific Coast. Horses carried salmon into the Great Plains. The Pacific Northwest tribes were traders and businessmen. The salmon was the economic base and also the foundation for their entire culture and existence. The Native Americans were the guardians and caretakers of the River. This tradition continues today.

In restoring their lost traditions, Native Americans practice the elements of the portfolio that we introduce in this book (see section 5–8). They integrate traditional ecological knowledge with western science. They make sustainable choices as summarized by John D Tovey in section 4.1.2. This is why it is worth mentioning several of their projects. These projects address the important cultural traditions that were impacted by the European colonialism and its battles to eliminate Native American traditions and culture (see section 2). Native Americans have not forgotten and become a blend of all the different cultures that migrated to North American. They have survived. They are restoring landscapes highly altered by past non-Indian land-uses. This is an important part of their culture—a culture that is not forgotten despite industrialization and the computer technologies available today. Tribal people do not avoid technology but are selective in what they incorporate into their practices.

Tribal projects to connect cultural and traditional values to natural resources are vibrant and part of nature. They are part of reservation life. Tribes do not need help in figuring out how to connect to nature. In this book, we present only two projects to demonstrate the type of cultural activities important to native peoples. Our small summary does not do justice to the numerous projects that tribes have designed, funded and employed tribal members to implement. This information needs to be the subject of another book. What we have included are representative examples from the PNW U.S. We encourage the reader to read more on this topic since they are fascinating stories of how a conquered people have become leaders in natural resource management in the Pacific Northwest U.S.

Tribes have used treaty rights to pursue environmental restoration and mitigation of their cultural resources. Without American tribes, salmon restoration would probably not be happening at such a large extent in the PNW U.S. Most cultural resources like salmon use large landscapes as their habitats. Tribes have adapted to the "boom and bust" cycles that even salmon experience. They share the bounty of the ocean with non-Indians even though salmon is a really important cultural icon for them.

Tribes continue to manage some of their customary lands that they used to own as a tribe. They pursue strategies and goals that will allow them to restore all lands to a more sustainable condition. Tribes can manage reservation lands but they are also attempting to restore customary lands that were severely degraded over the last several hundred years. Tribes have been using the Treaty of 1855 to re-establish their traditional knowledge to manage cultural resources in the PNW U.S. landscape. It is very unusual to have native peoples manage resources on lands that they do not control or own. Many regional benefits accrue to non-tribal communities and the business enterprises in the regional economies.

In most rural areas, tribes are important contributors to the local economy. For example, in 2010, the Coquille Indian Tribal government had a Eugene-based economic consulting firm measure what their economic impact was in the local economy. They only included The Mill Casino Hotel as well as the RV Park in this assessment. The results of this assessment were: "The study revealed that over 1,300 local jobs and over $125 million in economic output in 2010 can be directly or indirectly traced to business and governmental activity of the Coquille Indian Tribe provides." [1]

The Coquille Indian Tribe did not have vibrant resources-based business enterprises to develop their economies. They had lost their sovereign status in 1954 and only gained it back in 1989. When they were terminated, they lost their forest lands. It was not until 1996 that the state of Oregon gave forest land back to the Coquille Tribe. They were able to select from existing forestlands managed by the BLM. They did not own forestlands at this time.

Upon receiving the land, they began to restore the cultural resources and to manage the forests for revenue for the tribe itself. This was not an exploitative management or use of the forests. In fact, the Coquille Indian Tribe, a couple of University professors and the U.S. Bureau of Land Management have collaborated on how to design and implement the Northwest Forest Plan on tribal lands. They are following a very ecologically based management strategy that already looks more natural than the surrounding forests.

Tribes are local but also global sellers of natural resources. They balance managing resources and generating revenue from these resources so their harvest does not harm the local and regional environments. For example, the tribes in Washington grow dry land wheat on the reservation that is currently all shipped to Asia. The tribes are working at selling this wheat in the PNW U.S. instead of China so that the money can remain in Washington State.

American tribes are investing in the future of lands, water and natural resources in the PNW U.S. The Colville Confederated Tribes have 10 representatives that control water use because of the different jurisdictions that are involved (mining, forestry, and etc.). This is money that the tribe is placing towards mak-

1 //theworldlink.com/news/local/coquille-tribe-added-m-to-economy-in/article_a5c2d8f0-f874-5b00-b29b-86422344cc63.html, accessed 11 December, 2011.

ing better decisions collaboratively on local resources found on the reservation and on their customary lands.

Tribes do not want to consume resources in a similar manner as non-tribal peoples. They are not going to consume water in the same way and are not interested in following intensive farming practices in their regions. Other non-tribal farmers practice "artificial farming" because they have to pump water to the fields and this water pumping is heavily subsidized and makes no sense to tribes. The non-Indian farmers are not paying the full cost of the water they consume and do not see the damage they impose on the environment. They do not grasp the real water story because it is so heavily subsidized.

In the Columbia River basin, tribes have been the driving force behind the restoration of fisheries and forested lands along this river basin. Tribal members actively pursue projects to mitigate or restore lands that have been altered because of historical land-use practices implemented by Euro-American mangers. Native Americans in Washington State use traditional knowledge to make decisions regarding the use of their resources. When possible, they act in the best interest of "the seventh generation" after them. As an example, forest clear cutting has been eliminated under their stewardship along with the replenishing of fisheries. This standard has been applied to everything from forest management to salmon fishing. The individual tribes can fight amongst themselves, but when it comes to managing their resources, they work together for the overall good. This is the same story as the salmon chief (see Box 5).

Tribes take resource management very seriously. They are not interested in the intensive resource management. They use their sovereignty status to get the state and the federal government to uphold treaty agreements and implement environmental practices that threaten their cultural resources. Tribal collaboration on salmon and dams is briefly summarized next.

## 9.1 Salmon Restoration and Tribal Co-management

There is a history of treaty rights that give certain Columbia basin tribes rights to fish from their customary lands. These treaties also recognize customary treaty tribal rights to salmon and tribal peoples as co-managers of fishery resources. For example, Cosens wrote how the Article 3 of the Nez Perce Treaty reserves gives [125]:

> "[t]he exclusive right of taking fish in all the streams where running through or bordering said reservation is further secured to said

> Indians; as also the right of taking fish at all usual and accustomed places in common with citizens of the Territory." The language stating that the right is "in common with citizens of the Territory," was interpreted by Judge Boldt of the U.S. District Court, Washington in 1974, to entitle treaty tribes to up to 50% of the harvestable fish that pass (or would pass absent harvest en route...) the usual and accustomed fishing places. .. the ruling recognizing the legal right of Native American's equal access to fish ... the Ninth Circuit Court of Appeals interpreted the right of treaty tribes "in common with citizens of the Territory," as analogous to a co-tenancy, stating: [C]otenants stand in a fiduciary relationship one to the other. Each has the right to full enjoyment of the property, but must use it as a reasonable property owner. ... neither the treaty Indians nor the state on behalf of its citizens may permit the subject matter of these treaties to be destroyed."

Tribes are active collaborators on projects to restore salmon fisheries in the PNW U.S. The U.S. government guaranteed tribes sovereign rights to harvesting salmon in the 1855 treaties. However, tribal harvest of salmon is and continues to be threatened by dams that block their migration to spawn in the upper reaches of river systems. The degradation of the river systems is also further aggravating salmon populations.

Once treaty rights of Native Americans were recognized, several Northwest U.S. tribal governments united to restore the viability of salmon fisheries. Their goal is to be able to include salmon in their cultural practices. Most tribes are participating in the salmon fisheries recovery efforts. No matter what approach is taken, the goals are the same. Some tribes have formed partnerships among several tribes and others are pursuing these goals as a tribe.

The Lower Elwha Klallam Tribe advocated removing the Lower Elwha and Glines Dams. These dams were built nearly a century ago. The Tribe was the first to call for the restoration of the river and its salmon. In addition to advocating for dam removal, the Tribe has led habitat restoration efforts in the lower river, and operates a hatchery to maintain Elwha salmon runs. The Tribes efforts worked. The two dams were just removed a few months ago and salmon are already coming back up the tributaries of the river systems.

In another example, four tribal governments formed the Columbia River Intertribal Fish Commission (CRITFC) in 1977 to unite the efforts of the four tribal governments to renew their sovereign authority in fisheries management." [125]. CRITFC includes the Nez Perce, Confederated Bands of the Yakama Nation, Confederated Tribes of the Umatilla Indian Reservation, and Confederated Tribes of the Warm Springs Reservation. These four tribes historically "shared a regional economic economy based on salmon" [126].

"The tribes created a coordinating and technical organization to support their joint and individual exercise of sovereign authority. Based as it was on a time-tested tradition, the new organization, the Columbia River Inter-Tribal Fish Commission, became a valuable means for organized intertribal representation in regional planning, policy, and decision-making." [126]. Since CRITFC was formed, these four tribes have been co-managing salmon in the Columbia basin. CRITFC's mission statement is [123]:

> "The Columbia River Inter-Tribal Fish Commission's mission is to ensure a unified voice in the overall management of the fishery resources, and as managers, to protect reserved treaty rights through the exercise of the inherent sovereign powers of the tribes." [1]

In addition, five other tribes on the Columbia no longer have salmon migrating onto to their lands. These tribes formed the Upper Columbia United Tribes or UCUT. This group negotiated a memorandum of understanding with Bonneville Power Administration to ensure that their sovereign rights to salmon would be recognized [125].

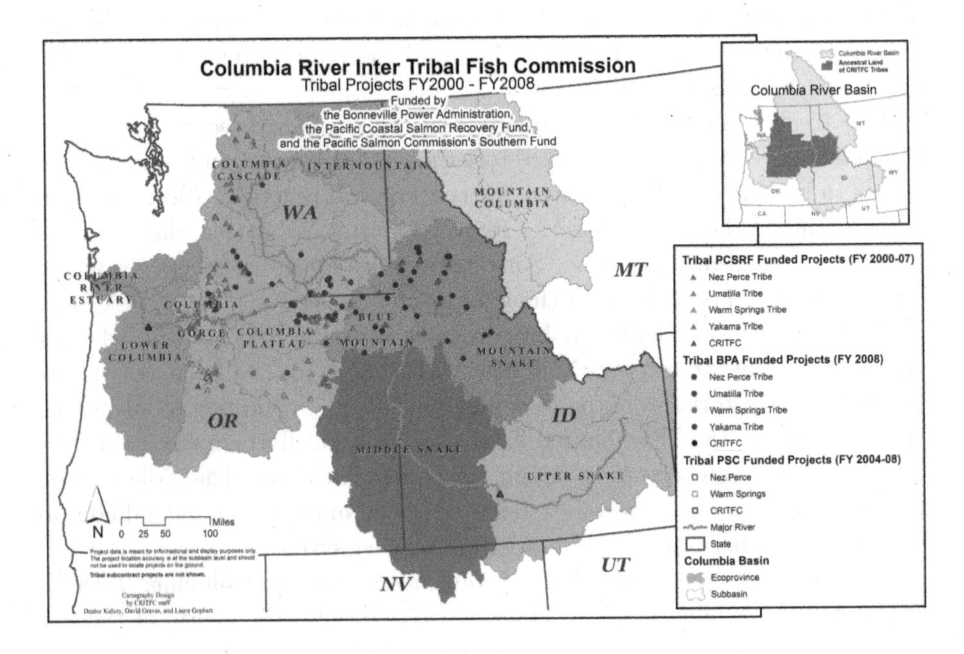


**Fig. 9.1** A map of the tribally managed projects on the Columbia River.
Figure courtesy: CRITFC www.critfc.org/.

1 www.critfc.org/text/work.html

A map of tribally managed projects on the Columbia River coordinated by the Northwest Indian Fisheries Commission (CRITFC) is shown in Figure 9.1 [126]. It reveals the extent of tribal involvement in the restoration of salmon habitat on the Columbia River.

These projects are unique in that many tribes (e.g., Nez Perce Tribe, Umatilla Tribe, Warm Springs Tribe, Yakama Tribe) through the CRITFC are partnering with non-tribal partners to restore habitats across multiple watersheds. Many of these projects are on ceded lands located in Oregon, Washington and Idaho.

These partnerships are also unique in that the traditional boundaries of management agencies are no longer relevant and do not hinder the restoration of large watersheds needed to restore viable salmon populations. Projects vary with 1) salmon habitat protection and restoration; 2) salmon research, monitoring and evaluation; 3) watershed and sub-basin planning and assessment; 4) salmon enhancement; 5) lamprey restoration; 6) wildlife enhancement, and 7) public outreach and education [126].

## 9.2 Dams—Removal, Mitigation and Redesign

The need for renewable energy development seems likely to be a major U.S. national policy goal for the next decade and probably beyond. There are some interesting trends in the hydropower industry. According the WDC Report [50],

> "The end of the 20th century saw the emergence of another trend relating to large dams—decommissioning dams that no longer serve a useful purpose... momentum for river restoration is accelerating in many countries, especially in the United States where nearly 500 dams have been decommissioned".

This has occurred mainly on the smaller dams. The larger dams, such as the Grand Coulee Dam, will be around for the foreseeable future. So tribes will continue with their efforts to mitigate the negative impacts from dams on the river systems.

Analysis of the hydropower experience also has other applications for other green renewable energy projects in general. The WCD Report notes that, "By the end of the 1990's, it was becoming clear that the cost of controversy could severely affect future projects for dams and stall efforts to finance other non-dam water and energy development projects" [50]. Analysis by the WCD Report commission showed "that 60% of the impacts were unanticipated prior to the project construction" [51].

Trust is a major issue and the track record of energy interests in assessing impacts or in dealing fairly with existing residents has not been very good in most past instances. An indication of this is that even the WCD Report points out the treatment of the Colville Tribes as a good example of non-equitable treatment of indigenous peoples. Despite this recognition, it took half a century to reach a monetary settlement for flooded lands on the Colville Reservation. The fishing economy, however, has yet to be replaced.

There is also an assumption that hydropower does not affect global warming, but this too needs further analysis. According to the WCD Report, "A first estimate suggests that gross emissions from reservoirs may account for between 1% and 28% of the global warming potential from GHG emissions... (and this) challenges the conventional wisdom that hydropower produces only positive atmospheric effects, such as reduction of emissions of carbon dioxide, nitrous oxide, sulfuric oxides, and particulates, when compared with power sources that burn fossil fuels" [50].

The WCD Report recommends that future development be guided by three policy documents: 1) the Universal Declaration of Human Rights; 2) the United Nations Declaration on the Right to Develop; and 3) the Rio Declaration on Environment and Development [50]. These policies highlight the linkages between environment and development but also acknowledge the importance of local communities having a significant role in shaping national development strategies [50]. The WCD Report also notes that there has been a shift in the definition of public interest, "from one that placed a premium on overriding interests of economic growth to one that places more weight on the rights and interests of people and communities affected by a development." [50]

In a nutshell, the World Dam Commission recommends that stakeholders be allowed to have meaningful participation in the planning and assessment of new projects. Projects should be assessed for impacts. There is a need for social justice, one group should not be put into poverty so that another group can be made wealthy. None of this was done in the case of the Grand Coulee Dam. Tribes were put onto the Colville Reservation, and then their means of making a living was taken away by the dam. Even a half century later when the Congress did make some compensation for flooded lands, there remain significant dam-caused impediments to economic development which still have not been addressed. In the meantime, generations of people have suffered poverty and have received inadequate assistance in rebuilding their economies, while other in the region enjoy the benefits from the hydropower.

Similar principles would apply to all other renewable energy projects also, though it is unlikely that many such projects would ever have the same magnitude of impacts as does the Grand Coulee Dam. Further mitigation and monitoring are also necessary for dams on the Columbia River and these efforts are ongoing. PNW U.S. tribes have been in constant negotiation and litigation to alleviate the impacts from the Columbia River dams since their inception. They will continue with this work. In 2009, several NW tribes negotiated an agreement with Bonneville Power Administration to implement nearly a billion dollars of dam mitigation projects to restore salmon on the river system.

The Columbia Basin Bulletin summarizes the results of this accord in Box 22.

**Box 22: Tribes, Federal Agencies Sign "Columbia Basin Fish Accords"**

The Columbia Basin Bulletin posted on Friday May 09, 2008 (PST)

"Four Columbia River basin tribes and three federal agencies say years of divisiveness over salmon recovery efforts in the Columbia River ended May 2 with the signing of agreements designed to deliver specific, scientifically valid biological benefits for the region's fish."

The signing of the "Columbia Basin Fish Accords" was celebrated with traditional tribal ceremonies…

…The agreements were reached as a follow-up of discussions that took place during a two-year "collaboration" ordered by the U.S. District Court of Oregon Judge James A. Redden… Redden required that the views of the of the states and tribes be considered.

…The agreements…were signed at… "Tsagaglalaiai" or "She Who Watches" Tribal legend explains that Tsagaglaiai was turned into stone so she could remain at her village to watch over the river and its people for eternTEhity…

"These accords move the focus away from gavel to gavel management and toward gravel to gravel management," said Steve Wright, BPA Administrator.

…The agreements were harshly criticized by the state of Oregon and by the fishing and conservation groups…

…The agreements guarantee nearly $1 billion in project funding over the next 10 years. They also make the pledge that the four tribes will not challenge in court or elsewhere…

(Agreements were with) the following entities:

- The Confederated Tribes of the Umatilla Indian Reservation
- The Confederated Tribes of the Warm Springs Reservation or Oregon
- The Confederated Tribes and Bands of the Yakama Nation
- The Confederated Tribes of the Colville Reservation

(Continued)

> Tribal leaders said the agreements represented a breakthrough in many respects...
>
> "We think these accords are a turning point in the way people in the basin address fish recovery and we look forward to being and integrated part of this strong partnership", said Mike Marchand, Chairman of the Confederated Tribes of the Colville Reservation. "This finally brings projects for fish in the upper Columbia. In the past, all production measures were put out of reach of the Colville people. These accords bring fish back to the Colville people."

On May 2, 2008, the Columbia Basin Fish Accords were signed on the banks of the Columbia River (see Fig. 9.2). These Columbia Accords were signed by Mike Marchand and R Sampson on a white doe skin. This BPA Agreement was for nearly a billion dollars for fisheries projects [127]. For specific details of the accords and the types of projects involved, go to www.critfc.org and www.salmonrecovery.gov.

**Fig. 9.2** Mike Marchand and R Sampson signing a white doe skin as part of the Bonneville Power Association Agreement.
Photo courtesy: BPA and given to Mike.

Together, leaders from the Colville, Umatilla, Warm Springs, and Yakama Tribes and agency officials from BPA and other federal agencies, gathered to sign a ceremonial deer hide to finalize a negotiated agreement.

This money would be used to make improvements on the dams to allow for better fish passage and also to improve streams for salmon. This work will be monitored and no doubt tribes will be back at the table to negotiate continued improvements in the next decade.

It was not a complete solution, but it was a first step in the long process to fix the dams. Prior to this, the tribes' only recourse was to take the agencies into court. A lot of money went to lawyers and solutions were not optimum. This negotiated settlement is hopefully a turning point, all the stakeholders working together need to achieve consensus on the needs of the Columbia River, the salmon, and the people. Hopefully this day was a turning point for the better [124].

**Coyote Essentials**

- Sustainable natural resource management has to be regionally controlled and framed in a local culture and traditions.
- Include all vested stakeholders in the management of natural resources and really collaborate.
- Never marginalize the up-river people that live and manage natural resources on their lands.

# Chapter 10

# Final Words on Essential Native American Leadership

Leadership cannot be just something that is learned from scholarly books or going to a class at a University. So how can western world scientists, decision-makers and natural resource managers become more American Indian in how they interaction with the environment? First, there has to be recognition that Native Americans do have something to contribute to resource management. There also has to be recognition that the process used by Native Americans to make decisions differs from the western world's scientific approach. To begin with native peoples have nature-based cultural traditions. It will be challenging for a western scientist to understand this approach since it is not based on large data-sets of facts and models. It cannot be done using fancy equations.

This might be easier to practice if people shifted from the idea of needing to force every immigrant to the U.S. to become part of a "melting pot". Leadership has to recognize that we all do not want to become the same and in fact we should not be the same. If it did occur then this would be a very boring world. When people migrate to another country, they should adapt to their new world but not necessarily give up all their culture and traditions that they bring to the new country. One reason that people should consider retaining their culture is that different cultures create diversity and diversity equals resilience. A resilient society is good because if it has a diverse culture then that society probably holds a much larger number of world views and ideas for innovation in which people can use to face adversity.

Native Americans have adapted to the conquerors but most have kept their culture. The idea of avoiding the conversion of all immigrants into a "melting pot" should be avoided and we suggest that the concept of a "salad bowl" should be considered when deciding whether or not to adopt a new culture while leaving the older cultures behind. The concept of a "salad bowl" is an important concept that frequently is ignored in the rush to improve people's economic status. So therefore this "salad bowl" concept is discussed in the next section.

## 10.1 "Melting Pot" versus "Salad Bowl Assimilation" Discussion[1]

About 20 years ago, my colleague John Karwowski shared an idea with me about immigrant assimilation in the United States. He was concerned about immigration and especially the changing attitudes about people who come from predominantly non-white countries as well as those who are non-Christians. There is a well-known metaphor that America is a "melting pot" of people from all over the world who immigrate in order to find a better way of life—safety from persecution, a job, a home, religious and other freedoms. The country was described as a fusion of various nationalities, ethnic groups and religious sects into one distinct people.

The process of an immigrant to become a citizen requires them to learn U.S. history, political principles, and civic customs, and to speak American English, to "blend in" as much as possible and, above all else, vow to uphold American laws. A pledge of allegiance to the country was imperative. People were expected to give up and deny their old ways, customs and often their traditional values. They were sometimes able to celebrate those customs through fraternal organizations, ethnic festivals, and, in private religious guilds or simply living in urban districts. These losses could often be very painful and greatly missed but the secular rewards were immense.

People often experienced great hardships to get here and the trade-offs were viewed as necessary and tolerable. They had to make their first loyalty to America and its ideals. Bruce Thorton, a research fellow at the Hoover Institution, opined that "If some custom, value, or belief of the old country conflicted with those core American values, then the old way had to be modified or discarded if the immigrant wanted to participate fully in American social, economic, and political life. The immigrant was the one who had to adjust; no one expected the majority culture to modify its values to accommodate the immigrant." This is the "melting pot" description of America and, in fact, what actually happened.

Many people also came to the new world especially in the immigrant waves of the late 19th and early 20th century. They came willingly, and even imparted on their children to disregard old customs, languages, accents, etc., so that they could better assimilate into the new world. This was also seen in Indian tribes where many people in the generation from the 1930's to the early 1980's were often passively Indians at best or outright denied any "Indianness" lest they be associated with poverty, and other social ills. Immigrants tended to try harder since they have given everything up to move to another land and pursue new opportunities.

---

1 By Melody Starya Mobley-Cherokee.

Perhaps we should consider another metaphor if we are going to have immigrants be productive members of society today and not lose their past cultures while merging into a new society. The metaphor we need to think about for America is the "*salad bowl*." In this theory immigrants would bring their native customs, values and beliefs (*salad ingredients*) to America and all of this would be united (*mixed*) to take advantage of the strengths and abilities that diversity yields. America would incorporate contributions that would make the country the very best that it could be and allow individuals to retain individual aspects of their countries of origin as long as they did not detract from the overall shared goals or the laws of the country. The United States would have optimized the benefits of full participation of all of its people, native and immigrant, in the development of its ideals and culture. Uniqueness and individuality in thinking and otherwise would not merely be tolerated at some level. All people and unique ideas that benefit nearly everyone would be prized, highly valued, encouraged and respected. This is my view of the "salad bowl" theory and how it could benefit others.

In a melting pot, the predominant, strongest flavored or loudest ingredient may be all that is perceived or tasted. If it does not contain a sufficient or adequate amount of another ingredient, that ingredient's delicious, interesting flavor may be completely lost or imperceptible. In a salad bowl, the overall flavor is enjoyed and each bite has a unique taste to enjoy because even minor ingredients are perceivable.

Of interest to me is the fact that American Indians were expected to assimilate in the same way as immigrants although they were here long before Europeans arrived and subsequently become the majority.

What Native American's bring to sustainable leadership is their culture and traditions. For example, one can learn culture and traditions by sitting on the knee of one's grandmother or grandfather. This approach provides the context that leadership needs. Memorizing leadership rules does not provide context. Rote memorization only gives you rules that you may not remember for when and where you should apply them. The next story is written by Mike Marchand about the lessons he learned from his grandfather who was the Chairman of the Colville Tribes.

## 10.2 Lessons from My Grandfather by Mike[1]

As a young boy, my Grandfather used to take me on tours of our lands. His photo taken in 1907 is shown in Figure 10.1—entitled Silico_Saska_wenatchee1907. My

1 Confederated Tribes of the Colville.

grandfather was a Chairman of the Colville Tribes (see Fig. 10.1). Our people used to have hereditary Chiefs, but in 1938, the tribe decided to have elected Councilman. Instead of a Chief we now have Chairman. Later, I would grow up and was a Chairman too. But as a young boy, we would jump in Grandfather's car and drive to the places important to our people and he would teach me about these places and about our people. One day he brought me to the mouth of the Entiat River where it enters the Columbia River. He stopped by the bridge and said that many years ago there was a village under the bridge, the village of the Entiat people. The site was now under water, it was flooded by one of the many hydroelectric dams on the Columbia. This dam was the Rocky Reach Dam.

**Fig. 10.1** Mike Marchand's grandfather.
Photo courtesy: Mike.

The Entiat tribe had been forcibly removed 128.7 km north, to the Colville Reservation in 1872, along with 11 other tribes. But our ancestors' lands were still important to us and we still return there today to pick berries and roots and also to fish and our ancestors are buried in these lands. Archaeologists say that people have lived in this area for over 10,000 years. Evidence of this has been found by the discovery of arrow heads, i.e., Clovis Points, made by these people thousands of years ago. These Clovis Points were found nearby at the Paine Field Airport in the town of Wenatchee in Washington.

Grandfather pointed to a spot and said right there used to be Chief Silcosasket's teepee. Silcosasket means Standing Cloud. Silcosasket was one of my Great Grandfathers too. I was wondering if the Chief's teepee was the best one in the village since he was the leader, but my Grandfather just laughed at me. He said actually the Chief's teepee was the poorest one in the whole village. He said the Chief gave everything he had to the elders and to the widows and to the orphans and to the poor. A man's place in the world was based on how generous he was and by how much he helped his people. Owning materials goods and riches and being greedy were not important things to the Entiat people. Chief Silcosasket was famous for being wise and honest. My Grandfather said that Indians from as far away as Oregon would travel up the River to Entiat to have Chief Silcosasket settle disputes. He would listen to both sides and make a decision. Indians would abide by his decisions. In those days being honest and fair was greatly admired.

Now 50 years later, I still think about this simple little story every day. I have been blessed to have been chosen to be a leader of my tribe for 7 terms on our Tribal Council. I still wonder every morning if my actions would measure up to my Great Grandfather Chief Silcosasket's expectations or whether my actions would be approved by my Grandfather the Chairman's expectations. My mother was also on the Tribal Council. Knowing the land, our people's history, and respecting the people and Mother Earth are important to the tribe. Our time on Mother Earth is short. But our Ancestor's spirits watch over us. Then we will have future generations to come and we must respect Mother Earth and take care of it so that our future grand children will have a place to live for all time.

We move now from cultural lessons passed down through the generations to examples of governance within some tribes. Creative governance from consensual partnerships is a hallmark of Native American practices. We describe this in the next section by focusing on one person, John McCoy (see section 6.2.2.2), and how he helped the Tulalip Tribe to develop their tribal economy. These are the practices that, if incorporated, could fundamentally change our western world's business plans. The question is whether we will be visionary enough to look at different models and gain from them while still retaining those items to be considered invaluable foundational elements from the current model.

## 10.3 Essential Tribal Leadership through Partnerships, Governance and Sovereignty

All elements of the Native American sustainability toolkit—maintaining their connection to nature, making practical and realistic decisions, following a Native American business model and creative governance from consensual flexible partnerships—were practiced by the Tulalip Tribes as they transformed and diversified their tribal economy. This is one of many successful Native American business stories where the Tulalip Tribes did not follow the western world model for business development. However they did connect their traditions with western world science and technology to reach their goal.

This transformation was helped by gaming revenues that then allowed the development of the Quil Ceda Village on tribal lands. Many Tulalip Reservation leaders played instrumental roles in building the Quil Ceda Village. "Quil Ceda's structure derives from the efforts of a number of people, including, perhaps most centrally, current Village Director John McCoy and Tulalip attorney Michael Taylor" (John McCoy is no longer the Village Director) [128]. We are focusing on John McCoy's story and his role in the building of the Quil Ceda Village for our book. Before we read John McCoy's story, a brief history of Tulalip Tribes is helpful to set the context.

Historically, tribes in the Tulalip reservation in western Washington U.S. were commercial fisher people. Gardner and Spilde [128] wrote in 2004 how "the main source of revenue for the Tulalip tribal government came from commercial fishing under the Treaty and from leasing reservation land to non-Indians." Maintaining a robust fishing industry as the core economic activity to develop the tribal economy was not realistic and tribal leadership recognized this. The 1974 Boldt decision gave tribes rights to harvest half of the salmon in Washington State and to co-manage salmon fisheries. But fishing is not the success story that will develop the future economy of the Tulalip Tribes. Fishing was already becoming less viable by the mid-1980s when "130 tribal members were licensed to fish,... 30 members licensed today" [128]. Fishing is still an important part of the cultural activities of the Tulalip Tribes even though it is not their main economic activity. Developing tribal economies optimally meant a diversification of tribal economies without shifting far away from their historical and cultural practices. How the Tulalip Tribe accomplished this is a story that is worth telling. The Tulalip Tribes provide an excellent case study of what can be done if essential leadership and adaptive leaders are given the authority to develop a new model of economic development.

The transformation of the Tulalip Tribes economic core from a fisheries focus to a diversity of economic opportunities is a story of success where this tribe balanced its sovereignty and cultural values with economic development. The first challenge the tribe faced was the need to design "an effective legal and political blueprint for the commercial center" that was to become the Quil Ceda Village [128]. This legal and political blueprint for how a sovereign tribe could collaborate with a state (i.e., Washington State) and retain their sovereignty as negotiated in treaties with the federal government did not exist at this time. The Tulalip Tribes had no blueprint or model that they could follow to build their commercial center. No U.S. tribe had attempted to develop and collaborate with a state government to diversify their economic base on tribal lands.

Forging collaborations with non-tribal entities and non-federal agencies to develop and diversify tribal economies appear to be simple on the surface. But it was not. Even though tribes have an abundant and diverse portfolio of resources on reservation lands, the non-Indian communities are acutely aware of, and want to have a say on, what happens on tribal lands. They feel that it will impact their revenue generation potential from natural resources and tribes would compete with them. Since tribes have rights to collect many customary resources from their ceded lands, conflicts have been the norm with non-Indians who are in businesses consuming these same resources and communities whose services are funded from resources collected from state trustee lands. Washington State's Department of Natural Resources (WA DNR) lands are former ceded lands that used to belong to tribes. These lands are now held in trust by Washington State for beneficiaries in the state, such as schools, Universities, and rural community services such as hospitals. Therefore WA DNR has a primary fiduciary responsibility to these other stakeholders who are not tribal. These other stakeholders maintain pressures on WA DNR to satisfy their trustee responsibilities since these revenues are used to support school and hospitals in rural non-tribal communities.

Resource conflicts became especially confrontational starting in the mid-1990s when the environmental organizations were successful in reducing the amount of timber to be cut (and subsequently reducing the amount of funds obtained from the timber harvest) from state forest lands held in trust and managed by the WA DNR. Trust lands are used to not only fund schools and Universities but fund repairs of buildings in the state capital, Olympia, and fund other critical needs of state institutions and local services provided at the county level. This funding model is a legacy of the fact that the U.S. federal government did not have enough money to fund state institutional structures and operations when the States first joined the union. Therefore, many western states that joined the union were given lands abundant in natural resources that they were supposed to sell to generate the funds they needed to operate. Today, rural communities continue to be directly impacted by any change to the level of funding provided from these trust lands since these funds pay for many of their basic services. One of the authors (Kristiina A Vogt) attended many DNR Board of Natural Resources meetings in 2000 where rural community or business representatives aggressively lobbied DNR to increase its harvesting of timber so DNR would obtain more funds to distribute to their trustee responsibilities.

This funding model creates an environment where it is difficult to create collaborative partnerships on how to manage natural resources. There are too many competitive demands on the same resources so someone has to be a loser. Gardner and Spilde [128] summarized well the conflicts that erupted between Indian and non-Indian fishers:

> "Tribal-state relations quickly deteriorated and acts of violence by non-Indian fishermen against Indian fishing camps erupted. Local press coverage only exacerbated the acrimony between non-Indians and Indians.... The contention leading up to the Boldt litigation and the frictions that followed the decision set a lasting combative tone for relations between many Indians and non-Indians and between some tribal governments and neighboring communities in the state."

Within the last year or two, WA DNR has been interacting with tribes and trying to determine how a state agency can collaborate with them to allow tribes to access resources from their ceded lands. Whether this is possible is questionable since the state has other fiduciary responsibilities to multiple stakeholders. This becomes especially problematic during cycles of economic downturn when rural communities are unable to maintain basic services in the community. They do not have other sources of funding to support communities facilities and institutions. The typical approach in the past was to take WA DNR into litigation instead of collaborating on how to "divide up" rich natural resources. This is part of the "boom and bust" cycles that are common in rural areas and where the economies are insufficiently diversified to adapt to the down-turns in these cycles.

Therefore building tribal business enterprises is a complicated problem where a tribe has to form partnerships with non-tribal entities that do not live on a reservation but are customers for resources and products produced by tribes. The state government is an especially important institution for the tribes to collaborate with because of their role in levying taxes and providing services at the county level. However, tribal interactions are government to government and tribal lands are held in trust by the federal government for the tribes. Tribes made and signed treaties with the federal government and not state governments. Therefore it is very unusual to partner with a state government.

Tribes have also faced a long history of non-Indian developers contesting any activities that tribes want to pursue even when they occurred on tribal lands. This placed tribes in a unique situation of "... contention concerning the development of natural resources within tribes, and between tribes and the state. Because of these diverging interests, the history of tribal-state relations in Washington has tended toward litigation rather than negotiation." [128]. According to Gardner and Spilde [128], a key turning point occurred when the Washington State government was willing to collaborate with tribes on the "passage of the Indian Gaming Regulatory Act (IGRA) in 1988"... since it did "force state governments into partnerships with tribal governments." As part of the Act, tribal governments were forced to negotiate a "compact" with state governments.

Of the 32 tribes in Washington, Tulalip Tribes is the only tribe that developed their commercial enterprises by forging a collaboration and partnership with the Washington State government to develop the Quil Ceda Village. It is a story that would not have happened if any of the collaborators would have been unwilling to cross artificial and real boundaries. It required a healthy dose of trust by the tribes since historical interactions had been decidedly one-sided and where mostly tribes lost land, lost their customary lands as well as the ability to make their own decisions (see section 2).

For the Tulalip Tribes, building a collaborative partnership with the state was possible because of tribal leaders who were willing to forge deals with non-tribal entities as long as they did not impact tribal sovereignty.

Leaders in this context have to be able to balance the western world demands without threatening Tulalip Tribes sovereignty. This is the story of one of those leaders—John McCoy—who not only has been a tribal leader but holds a legislative position in Washington State's 38th District (see section 6.2.2.2). Several other leaders were also important in providing leadership but we are focusing on his story.

### 10.3.1 One Tribal Business Model: Tulalip Tribes Building A Federal City[1]

As recounted by John, the Tulalip story starts with a vision held by tribal elders since the 1960s that something was going to happen on the north border of the Tulalip reservation. Their vision was the building of a big trading post. This foresight by the elders was visionary since the Quil Ceda Village now exists on this land and it is a successful modern day trading post with "outlet malls", a casino, gas station and restaurants. This of course is getting ahead of the story since this land was occupied and leased by Boeing Aerospace Industries before the Village was built. A trading post could not be built on this land as long as Boeing had a lease to it. The tribal Board wanted Boeing to leave the reservation land before their 2001 lease expired; Boeing also had a 10-year lease option after 2001 on this same piece of land. This task of negotiating with Boeing was given to John who had only been on the job for two weeks after leaving Unisys (see section 6.2.2.2). He was also asked to negotiate with Boeing to leave this land two years earlier than the expiration date of the Boeing lease.

John started negotiating with Boeing but quickly realized that he needed to make transition back to being an Indian if he was going to speak for the tribes and to present the tribal case to non-Indians. His 20-plus years in the U.S. Air Force, and working for computer industries outside of the reservation, gave John the knowledge and tools for how to speak like a white person. But, now he needed to be able to speak Indian to talk to tribal members while speaking like a "white man" to Boeing. Interestingly he needed to *relearn* how to speak Indian as well as to change how he dressed. The biggest challenge for John was to work on the "language" to speak like an Indian. John had help from his cousin to learn to speak Indian again. Every time John used a word that his cousin did not understand, he would throw a nickel at him. No matter where they were, a nickel would fly through the air and hit him when John spoke a word that his cousin did not understand. John learned Indian again quickly so nickels weren't thrown at him. He succeeded in thinking about his vocabulary and how to speak white *and* how to speak Indian.

---

1 As told by John McCoy, Tulalip Tribes.

Now John was able to talk to the tribe but also to the federal government and also the state that infringed on tribal sovereignty and Boeing. Next he figured out that it was better not to go to the courts if he could resolve the problems before that stage. At this point, John decided that it was better to lobby Olympia, Washington (the capital of Washington State) instead of dealing with these issues in the courts. While lobbying Olympia legislators, John needed to deal with reservation projects to sustain the Tulalip business enterprises since that was his job. John also recognized that it was very important to keep non-tribal impacts from derailing tribal business projects.

The key to all of this was to make sure that whatever activity occurred always flowed from tribal sovereignty to tribal economic development. As long as sovereignty was satisfied, the economic development activities worked for the tribe. Since the state- and county-level connections could destabilize economic development of the tribe, it was critical to form for the tribe different sustainable governance and economic development models that were compatible with non-tribal governance.

At the same time it was important to develop the human capacity side of the tribe. He needed to transform tribal business practices. Documents were written that established the rules for how business would be conducted that differed from the traditional tribal business practices. Now, management would not come down on you if you played by the rules that were written down. Before this time, the tribe had too many rules that were unwritten so the power rested with the employee. By rewriting the human resources manual and getting the tribal approval of these new rules, it established how governance would be implemented. Many tribal members were not happy with these changes and wanted to terminate John's contract. But he survived and his ideas survived. These rules made it where you showed up for work and you had to produce when you were working. This was a very different environment since the Tulalip's prior work had been in fishing which doesn't work like a business. It was no longer an issue of just knowing the tides for fishing but now in this new economic business many other factors were involved in judging the quality of a person's work. This required tribal members to shift to a very different work ethic. It was a big cultural shock for some tribal members.

In trying to develop a "federal city" many different problems had to be solved, some of which weren't obvious at first. For example, it was not very clear whether the state or the tribe had regulatory control on activities (legal and illegal) that occurred on tribal lands. When conflicts occurred between tribal and non-tribal organizations, who decided what actions were legal and who had jurisdiction over these activities? Were non-tribal organizations the decision-makers or did the tribes have rights on deciding the punishment of tribal members who broke the laws? An important question to settle immediately was when non-tribal organizations have the purvue over the tribe to judge and evaluate the legal ramifications of tribal actions on tribal lands. No business could be successful with regulatory uncertainty. Who had the legal right to decide what happened on tribal lands?

Who had sovereignty when tribal members broke the law—was it tribal courts or the non-tribal courts? These questions arose to forefront of the legal debates when one day John was driving his car on the reservation and did not come to a complete stop at a stop sign. A Snohomish county sheriff happened to be at the scene and saw that John had not come to a complete stop before driving on. The sheriff gave him a traffic ticket for not stopping. John was going to write a check and pay the ticket at a non-tribal court. However, Mike Taylor—Tulalip attorney—called him and told him not to do anything until he said so. John wondered how he even knew that he had gotten a ticket but Mike did not tell him. Mike told John in that conversation that he finally had a case that he could take to the courts to legally determine who had rights and jurisdiction on what happened on tribal lands. This is when John was introduced to *Public Law 280*. Mike won the case and the court decided that John's infraction was a civil offense and should go to the tribal court and not the Snohomish county court. This clarified the sovereignty rights of tribes and who had jurisdiction over tribes when legal infractions occurred.

Next John needed to figure out how to build the community infrastructure that was essential for building businesses on the reservation. This was an expensive option but something that was needed if any businesses would have a realistic chance of succeeding. After 1996–1997, the Tulalip Tribes recognized the need to upgrade their phone system on the reservation. John forged collaborative partners between educational institutions and the Tulalip Tribes to assess and design what was needed on the reservation. The internet was not available on the reservation but it would be indispensable if the tribe was to develop its economies. However John was told that it was not economical to develop tribal communication infrastructures. Instead of John accepting this statement, he decided to forge novel and new collaborations with community colleges to have students in class projects assess and design the communication systems and its implementation. John approached Everett Community College in 1999 and asked students how to bring technology to the reservation. At the Everett Community College in a nearby city of Everett, Washington, electrical engineers and computer science majors designed the technology that would function for the reservation's new requirements. The students wrote the concept paper. John then included University of Washington's (UW) Bothell campus (another nearby university) as part of the proposal writing activities. All these collaborations helped develop a proposal for moving their business plan forward.

Tribal electives heard about the student plans on how to automate reservation communication systems and how much savings would accrue from following this plan. The tribal Board approved the proposed plan and provided $300,000 to conduct a more thorough assessment of the plan. This assessment was done by the UW Bothell and Everett Community College students. The tribe opened its doors to these students who roamed all over the reservation during their assessment.

As part of this strategy, every five weeks a meeting was held at UW Bothell to give a debriefing. This allowed suggestions to be made and course corrections initiated. During this time, John realized how serious the students were in providing this feedback and the strength of the tribal and non-tribal collaborations that had formed. At that time, John found out that he had a mannerism that he was not aware of. As the students were presenting their briefing, John would ask a lot of questions. John was told later that the students had observed that whenever he was going to ask a question or make a comment, he would run his hand across his face (this was a trait that John had never noticed himself). One day, a female student that was being debriefed on her work by John fell to the floor because she was distressed when she had noticed that John had run his hand across his face.

These students were important in laying the foundation for implementing the programs to get fiber optics installed on the reservation lands. They designed strategies to get the communication network set up on the Tulalip reservation that were synergistic with non-tribal collaborators. This is an excellent example of consensual governance where a tribe did not pursue solutions in isolation from non-tribal entities. Today the entire reservation and the Village have their own telephone company and internet company who manage communications on the reservation.

Now we have to return to the Boeing story and finish how Quil Ceda Village was built. John was instructed by the Board to get Boeing to give up their lease on this northern corner of the tribal lands that would ultimately become the Quil Ceda Village. The Army originally had the lease on Tulalip reservation but this lease was transferred to Boeing by the Army. There is no documentation of why this transfer occurred when the Army gave up their lease. These lands were important testing sites for Boeing. They tested the damage that would occur when birds flew into airplane cockpit windows. They would shoot chickens into the windshields of aircraft to test what kind of damage would occur to airplanes at this impact. They also tested how the three stages of the rocket boosters separated before they sent John Glenn into space. When Boeing bought McDonnell-Douglass, all the negotiations that John had worked on were put on hold. However Boeing finally gave up its lease and even left two years before the lease was to expire.

Once Boeing left, the tribe needed to figure out what to do with this land. Mike Taylor came up with the idea of converting this land into a federal city just like Washington DC. This is what happened. As summarized by Valerie Red-Horse [129],

> "But what has made Tulalip such an outstanding role model and what draws other tribal leaders from all over the nation to tour and study for their own ideas, is the development known as Quil Ceda Village. The Tribal Government Tax Status Act of 1982 allows Indian tribes to create politically separate subdivisions, and the tribe incorporated the village in 2001 as a city directly under the U.S. government as a political subdivision of the Tulalip Tribes, with its own charter and ordinances. As far as the tribe knows, no other tribes have taken advantage of the law, and Quil Ceda Village is the only federal city in the nation other than Washington, D.C. It consists of 2,000 acres designated for economic development and was funded and developed 100 percent by the Tulalip Tribes utilizing their gaming proceeds to further additional economic development."

Today there are only two federal cities—Washington DC and the Quil Ceda Village. The Village structure allowed the Tulalip Tribes to give leases to non-tribal businesses. It defined the permitting process and made it clear what needed to be done by the tribe. The only problem that arose was that any activity occurring in this Village would be under the jurisdiction of the tribal courts. Some of the non-tribal businesses did not originally like this but it has not created a problem for the Village.

The Tulalip Tribes still needed to figure out how their federal city or Village could produce a tax base that would support its activities. For this Village to be sustainable, it needed its own tax base according to John. Students identified that any economic enterprise needed to satisfy certain factors for it to work for a tribe:

- It has to be on trust land.
- It could not do anything under the authority of the state and had to be federal.
- It had to be run by the tribes.

The Tulalip Tribes to date has not been able to produce a tax base from the Village. The State and County are able to do a lease hold tax on assets in the buildings. The tribes said that they should not be able to do this since it is on tribal lands. The tribes are still not able to garner any portion of the state and county tax on businesses on tribal lands. The tax situation may finally be changing since the end of November 2012, the federal government made a tax decision that is beneficial to the tribes. The decision preempts the county and state from assessing a lease hold tax in lieu of a property tax on tribal trust lands. This decision may finally allow the Tulalip Tribes to produce a tax base from the Village.

The tribe committed to building the supportive infrastructure needed by the businesses located at the Village which they have provided. Despite not acquiring a tax base from the Village, there have been many other benefits of building the Village for both the tribes and the adjacent non-tribal communities. This part of Washington has now become an attractive tourism and shopping destination because of what tribes built on this land. It supports the local economy [128]. It has also provided considerable employment opportunities for tribal members, mostly in the higher paying jobs found in the casino, so it is helping to alleviate the high unemployment found on Tulalip reservation.

The development of the communication network on Tulalip reservation and the building of the Village happened because of tribal leaders who practiced the rules that we have articulated in our portfolio. Without these practices, the Tulalip Tribes would have found it more difficult to develop their economies. The communication network was needed to support the business enterprises that became part of the Village. It also required leaders who were "renaissance" leaders—as defined by the old European idea of someone who was able to think out of their cultural framework, to have knowledge of all of the classics that make you a highly civilized and a cultural person, and able to connect and speak in two different worlds. It needs leaders to be able to think out of the box as the saying goes. It also requires people who are willing to interact with those who are not in their cultural group. It needs progressive tribal leaders who are able to recognize the past but do not focus on a vendetta that "throws out the baby in the bath water" without acknowledging that all people have something to contribute to societal resilience. It also requires non-tribal people to be willing to talk to tribal people. This type of person recognizes that it is okay to make mistakes but not to repeat mistakes (i.e., adaptive). It takes progressive leaders who surround themselves with smart people and then manage them. It won't work if the leader is a type of person who thinks they are the only smart persons in the room. It also does not work well if the decision process becomes bureaucratic and therefore non-adaptive. This requires people who do not give up their cultural norms but are able to balance their cultural roots with those of other societies. This allows them to adopt and adapt to a world that is dynamic and volatile.

This is where John McCoy excelled. His ability to operate in both worlds and to solve problems in both worlds was crucial. He recognized that many of the legislative bills being worked on in the Washington State legislature would impact tribes. By being elected to his Washington State legislative position, he has been able to educate his fellow legislators on tribal sovereignty and suggests solutions when problems arise. He also is able to balance the demand from the tribes and non-tribal stakeholders so that no one side dominates how he makes decisions. He has looked for solutions in both worlds instead of relying on his viewpoint from just one world.

Many of the U.S. land management agencies and non-profits have been partnering with Native American tribes to pursue their management goals. This acknowledges and recognizes that different values and practices need to be included in land management. It also suggests that the principles that we have articulated here are recognized for their validity in resource management. This is a good sign even though more of it is needed.

### 10.3.2 Increasing Collaboration on Nature Using the Native American Approach[1]

Increasing collaboration among government (federal, state, and local) agencies, Native American Tribes and Alaska Native Corporations, and private organizations (including non-governmental organizations and non-profit organizations) reflects the new approach being taken by the western world towards natural resources and nature. The increase in collaborative projects involving Native American tribes and natural resource management agencies in the United States reflects emergent trends:

- Use of collaborative approaches between agencies and groups in managing natural resources is a priority based on recent Executive Orders, national direction, and court decisions;
- The concurrent increased recognition of Native American rights and the need to work with them as sovereign governments with rights of self-determination;
- Consultation protocols and processes are becoming better defined, and more standardized and formalized within agencies;
- Agencies and Native American tribes take advantage of efficiencies that result from collaborating in natural resource management to achieve mutually desired objectives; and
- Agencies learn about traditional ecological knowledge (TEK) from Native Americans and Native American tribes have more access to Western science and procedures.

Each of them benefits from the sharing and transfer of knowledge, scientific and cultural. As budgets and staff continue to decline for agencies and tribes, natural resource managers gain efficiencies by using various combinations and levels of sharing procedural knowledge and knowledge of the natural resources, funding projects, sharing staff, and assigning responsibilities on specific projects. Collaboration with tribal leaders can help agencies mitigate contro-

1 By Melody Starya Mobley-Cherokee.

versy and contentious management issues. Collaboration helps build ownership in the management of public lands and in specific projects. Increased use of collaborative processes in federal resource management, combined with legislative and socio-cultural developments pertaining to tribal sovereignty and culture, has contributed to an expansion of collaborative arrangements between Native American tribes and natural resource management and other agencies.

For much of the period since the U.S. government and tribal governments signed treaties as independent sovereign entities, concern for trust responsibilities and treaty rights has been eclipsed by the prevailing objectives to assimilate American Indians and terminate tribes. Since the 1960s, however, several pieces of legislation and executive directives have been enacted that were designed to protect the rights of tribes and create a legal framework for collaboration [e.g., President Clinton's 1996 Executive Order 13007, the National Indian Forest Resources Management Act of 1990 (Pub. Law 101-630 Title Ⅲ, 104 Stat. 4532) and the Tribal Forest Lands Protection Act of 2005 (118 Stat. 868-871, 25 U.S.C. 3115-3115a)].

Also, several decisions by the United States federal courts have reaffirmed the sovereignty of tribes and outlined their roles as co-managers of treaty-protected resources, such as the 1966 Belloni Decision on tribal treaty fishing rights in the Columbia Basin (302 F. Supp. 899), the 1974 Boldt Decision on tribal fishing rights in Washington (384 F. Supp. 401), and the 1983 Voigt Decision pertaining to treaty rights of the Wisconsin Chippewa (700 F.2d 365). With these enactments, formal dialogue and engagement processes between American Indian tribes and the U.S. government have been mandated to protect tribal treaty rights, facilitate agency protection of tribal interests, and promote agency consultation and coordination with tribes.

The executive and legislative developments demonstrate an evolution from discretionary considerations of tribal interests in federal projects to mandated government-to-government consultation and inclusion. Presently, federal land management agencies must consult with tribes where: tribal rights are reserved by treaty, spiritual and cultural values and practices exist, public lands are adjacent to tribal or trust lands, and tribal water rights may be affected. These developments demonstrate that even with over a hundred-year history of treaty rights, collaboration between American Indians and the U.S. government in natural resource management is relatively new.

Numerous factors affect collaboration between agencies and non-governmental entities. Barriers to collaboration include power differentials among stakeholders, unclear or inflexible legal authorities and administrative policies; organizational cultures unaccustomed to collaborative processes, agency fears of losing control, and funding availability. Factors which promote collaboration include shared and open decision-making processes, goal-setting early on in the process, and continual information sharing. In addition, stakeholders' willingness to share authority and benefits, provide resources, acknowledge the legitimacy of other stakeholders, be flexible, and trust other stakeholders increases the likelihood of successful collaborative arrangements.

The emphasis on empowerment of local residents and communities by treating them as equal participants in resource management decision-making has its roots in many developing countries where participatory development has been attempted and espoused for several decades [130, 131]. Agency directives for collaboration provide institutional backing and programmatic structure to collaborative processes [132, 133], but procedural flexibility has been identified as being important as well [133–135].

Collaborative arrangements that identify a common goal and define stakeholders' roles and responsibilities contribute to successful collaboration. Institutionalizing the collaborative process through a variety of mechanisms, such as contracts, partnership agreements, and memoranda of understanding contributes to sustainable collaborations because they provide structure and validity [136]. However, collaboration is also understood to be an evolving process involving social learning and flexibility [137].

Rigid institutional arrangements can become obstructive to the adaptive nature of collaboration, leading to calls for policy reform [138]. How different institutional arrangements and mechanisms influence the social dynamics and structure of collaborative processes and thereby shape their form and function is not well understood. Many definitions of collaborative resource management exist.

Since the 1980s, natural resource managers around the world have looked to indigenous groups and their knowledge to manage processes and functions of complex ecosystems [139]. Although many definitions of traditional ecological knowledge (TEK) exist, it is generally considered the "cumulative body of knowledge, practice, and belief, evolving by adaptive processes and handed down through generations by cultural transmission, about the relationship of living beings (including humans) with one another and with their environment" [139]. Knowledge development is based on detailed observation of the natural environment, feedback learning, links between society and the environment, and resilience to changes within the environment. Traditional ecological knowledge is often considered as being ethically-based, spiritual, intuitive and holistic.

By contrast, western science tends to focus on understanding small parts of larger systems that separate humans from the natural environment (see section 8). Western science combines a particular set of values with systems of knowing based on empirical observations, rationality, and logic as opposed to perceived truths or perceptions. According to Kimmerer [140], incorporating TEK into natural resource management practices is one way to validate and include tribal abilities. Using the above definitions as a guide, TEK may be viewed as a process that incorporates tribal culture, values, practices and beliefs, as well as the relationships that exist between humans and the natural environment.

Application of TEK in research and management of public resources raises several concerns. Once written, codified, or taken outside of its cultural context and put into another frame of reference, TEK can assume different meanings [140]. Moreover, integrating TEK with western science imposes non-native ideals about knowledge and life experiences of native people and forces researchers to compartmentalize and distill indigenous beliefs, values, and experiences according to non-native criteria [141]. Finally, TEK varies within and between individual tribes and will not be practiced in the same manner by each community (see section 5).

Many governmental and non-governmental agencies have already recognized the importance of what Native American tribes contribute to natural resource management. This trajectory of resource management agencies regarding the inclusion of the Native American behavior is exactly what we have described in this book. It shows that our toolkit is a viable approach that needs to become more commonly part of our global sustainable practices. We are documenting the increasing collaboration that is occurring among government and private organizations. This change has taken a long time to emerge in the western world, especially after several hundreds of years of battling Native Americans and forcing a western world view of what it means to be civilized.

We think that the Native American model of keeping one foot in the past while making decisions that factors in "7th generations into the future" can begin the trajectory for industrialized societies to begin the journey towards ecosystem-based decision-making and essential sustainability. The key to the Native American model is their continued connection to nature and where decisions on resource uses always include, not exclude, nature. Native Americans also set aside certain culturally-based factors that are not part of the tradable or negotiable items. This decreases the potential for a special interest group to control the process and negotiate away something with greater value to the group.

We are not suggesting that using traditional knowledge is the only approach to make sustainable decisions. We are saying that the *process* that Native Americans follow is important to emulate, especially the process they follow to exclude some resources or land from being included in the negotiation process. Native Americans use traditional systems-based knowledge in their decision process but also are superb adaptive managers. They are very facile in developing totally new business enterprises and readily adopt technology when it clearly does not jeopardize their cultural resources. They do not glue or anchor themselves to traditional ideas or approaches. They are open to new ideas as long as valuable cultural resources are not included in the negotiation process and potentially lost from the community. We contend that traditional ecological and cultural knowledge held by many indigenous communities provides an approach to humanize society's resource decisions. It also allows societies to make locally-based decisions. *Traditional knowledge does not provide specific solutions for a specific problem but is an approach for humans to follow that specifically includes nature and humanizes the decision process.* Now let's go back to our story again.

## 10.4 Essential Sustainability: Building A Native American Behavior and Thinking Toolkit

There are many models that can help us change how we make decisions. Ours is a simple model: *Become an American Indian*! Or at least begin to think and behave like an American Indian. This will make western world trained decision-makers superb adaptive *managers*. Native Americans and other indigenous groups have less difficulty in dealing with the sustainability concept because of their traditional knowledge. These are the people we should look to when figuring out whether we are on the right track for making sustainable choices. Becoming an Indian does not mean you should look cosmetically like an Indian but it means you should adopt aspects of their political, cultural, spiritual or religious and economic characteristics. It also means that you need to foster a local-based understanding and centric world view of nature.

Most indigenous communities in general have very different sustainability practices compared to western Europeans. Until quite recently, most indigenous people's concepts have to be translated through western concepts in order to be "understood" by western societies [69]. Much is lost when concepts have to be translated using another language. The value of our book is that we are not trying to translate indigenous concepts only using the western world language. We show where Native American and western world practices need to connect and become compatible. No Native American would suggest the western world practices and technologies should be thrown out and not used. They do say that the western world practices are not in balance with the "human" side of society (see section 1).

Indigenous communities have developed a history of cooperation in resource management with an emphasis on multi-tiered environmental governance. These approaches need to be understood and not altered or repackaged using someone else's language. It is their core values and traditions that are the basis of these different models espoused by indigenous communities. It has worked exceptionally well in conservation which has been problematic for western scientists. This has been less of a problem for native communities where [21]

> "The close connection between culture, nature, and survival that led to native conservation practices and designation of sacred places has been

hypothesized as enabling many indigenous cultures to support high levels of natural biodiversity in their homelands for many generations".

Indigenous communities have evolved and refined their practices over several hundred years. In general indigenous communities share several similar characteristics with Native Americans:

- Rely on their mythologies and traditional knowledge to maintain and promote a long-term view.
- View everything as interconnected so there is a need to consider everything in coupled social and natural ecosystems.
- Look at the Earth as alive, whole and having a "soul", with no physically constructed boundaries. Tribes do not build boundaries but use natural features in the landscape; e.g., John D Tovey's grandmother told him stories about how the Tucannan River north of Walla Walla was one of our historical boundaries of his tribe.
- Tribes view nature as a whole and where everything has a "soul" from a person to a rock. With this approach, being a vegetarian is a funny concept since each plant has a "soul".

These attributes of indigenous communities made them adept as ecosystem and adaptive managers before these concepts became popular in the western world. It is amazing how similar the Native American attributes are to the core principles of ecosystem management developed by the western world only in the early 1990s [74, 142]:

- Focus on the sustainability of ecosystems, not on the output of products;
- Adopt a holistic understanding of the way all the parts are linked together in an ecosystem and the feedbacks among those linkages;
- Incorporate a long-term perspective and examine issues at a scale relevant to the functioning of the ecosystem; and
- Recognize that human values shape ecosystem structure and function in myriad ways that can constrain, promote, or reduce resilience and vice-versa.

These similarities suggest that the western world does not need to develop new paradigms for sustainable practices. We can learn from Native American practices that have been around for several thousand years. Their practices already integrate all four principles of ecosystem management mentioned in the above paragraph. Tribes and indigenous communities were the Ultimate Ecosystem and Adaptive Managers before it became the "fad" to espouse these practices. The scientific approach to ecosystem management continues to be "to manage values that we want out of our ecosystems". Most Native American tribes have stories or similar folklore that can provide considerable insights into their dec- ision process when compared to formal western science stories, e.g., a peer reviewed publication.

Even though the western world scientists did develop indicators and rules for ecosystem management, they mostly have not connected the four principles simultaneously to address a problem. The western scientific approach is excellent at developing new paradigms or principles but frequently this occurs at the expense of including all necessary components needed for a system to be adaptive. A new name or paradigm doesn't mean that we are practicing better management if we do not include all principle components essential for retaining the functions or structures of an ecosystem. A good analogy for the first principle of ecosystem management is the focus on a product output instead of sustainability of the ecosystem. If a cook makes a chicken dish but "this dish does not include the main ingredient—a chicken—but has snake meat or tofu substitutes", the product may look and taste like chicken even though it is not a chicken dish even if you call it by that name.

History or traditional knowledge helps us understand what we should include when practicing traditional ecological management. Traditional practitioners would never substitute snake meat into a dish and then tell you that you are eating chicken. They would not focus on the output of products but how a decision would change the connections in an ecosystem. The same cannot be said for a person who is part of a stakeholder group that would benefit economically if they managed to get you to eat the false chicken dish. They may have just gotten a really good deal on several tons of snake meat after a snake round-up. There probably was no question asked about all the snakes that had been killed and whether it was detrimental to the health of the desert ecosystem. The economically benefiting stakeholder group probably also considered it is best to focus attention on only the product and not on whether the ecosystem becomes less resilient.

Therefore a leader should be a real creative and systems-based decision maker. Having this quality is important since business, societies and the environment are very volatile. They change but not at the same time. This creates conflict and aggravates resource scarcity. There will always be some places in the world where resources are not scarce. Other locations may not be so fortunate and face scarcity of resources. Since nature provides resources and ecosystem services in "boom and bust" cycles, any toolkit has to provide leadership for these changing conditions. This means that decisions are not only made during good times but also the bad times—this is probably why the coyote has elements of both. This is similar to a marriage contract where two people promise to respect and love one another through all the good *and* bad days. Unfortunately, it is not uncommon when the times are bad that marriages break up. People make a vow but some people cannot deal with the hard or "bust" times. A real leader will need to provide leadership during both the "boom and bust" cycles. Discussions of how to do this follow in the next section.

## 10.5 How to Do Business in A "Boom and Bust" Economy

Boom and bust economies are triggered by natural and human disturbances. When humans produce "boom and bust" cycles of resource supplies, these are artificial collapses in resource supplies. Recently, corporations and governments are "grabbing land" in other countries because they do not have enough productive lands to grow their food or energy crops [143]. The driver is concern over energy and food security [143]. These "land grabs" artificially cause a "bust" in the economy where it plays out. It continues the thread of the up-river and down-river metaphor that we introduced in section 1. The people who live on the land are the ones that usually lose out when these land grabs occur. They are the ones who are removed from their lands as foreigners acquire "better uses" and mostly "more intensive" uses from these same lands. This was recently described as [143],

> "...some corporations and governments are investing in agricultural land as part of a long-term strategy for food and energy security... number of land-related deals has dramatically increased since 2005, reaching a peak in 2009 (8). In 2010 the World Bank estimated that about 45 million ha had been acquired since 2008; most of these land deals were for areas ranging between 10,000 and 200,000 ha (9). Moreover, several institutions [e.g., the World Bank (WB), the Food and Agriculture Organization (FAO), and the International Fund for Agricultural Development (IFAD)] have reported that many deals were closed with limited consultation of the local population, without adequate compensation of the previous land users, and without seeking opportunities to create new jobs or enhance environment sustainability."

In contrast to these artificially created "busts", most indigenous communities are adapted to "boom and bust" cycles and carefully make decisions that modulate the "bust" part of the cycle. Tribal businesses and rural resource extractors are superb "boom and bust" cyclers. They have traditional ecological knowledge that goes back some "7 generations". They have adapted to the fluctuations in resources caused by weather events. Nature has cycles where periods of resource abundance will be followed by scarcity. This is the normal situation. This is why we end this book by mentioning the "boom and bust" cycles since we need to design our businesses and human development opportunities to be adaptable to these cycles. We rarely face a situation of hundreds of years of resource abundance.

A successful business venture has to be able to operate through the "low and the high" of any cycle to survive. The recent article published in Science Now [144] shows how Alaskan Aleuts dealt with a "boom and bust" cycle for their food supplies. They recognized the cycles of nature's abundance and their practices kept their ecosystem resilient, increased the diversity of food webs and kept parts of the food chain from collapsing. As Travis [144] writes,

> "For most of the past 5000 years, the hunter-gatherers known as the Aleuts have lived on Sanak Island off the southern coast of Alaska, surviving on the local fish and marine mammals they caught as well as clams and mussels collected in the intertidal zone around their island. Now after constructing some of the most elaborate food webs ever built, a research team has begun to reveal how the Aleuts fit in as a top predator in the island ecosystem. In short, the Aleuts weren't picky eaters, consuming about one-quarter of the different species on and around Sanak Island. But by being such "supergeneralists", the research team suggests, they were likely able to keep the ecosystem stable because they would switch prey when a particular species became endangered, and thus harder to catch or collect."

The Alaska Aleuts practice survival in a landscape where natural resources, i.e., food, can be abundant or scarce [144]. They would not have survived in Alaska for so long had they not adapted to their environments.

Businesses need to think about what the Alaska Aleuts practice. Business conditions and practices are not stable. Stability is not the norm. If we tried to stop the "change" and to eliminate the "boom and bust" cycles that are part of life, entropy would rule.

Taleb [80], a former derivatives trader, summarizes well in his new book (*Antifragile: Things That Gain From Disorder*) how "unexpected events" occur in an unpredictable frequency. He writes that humans need to "learn to benefit from disorder" and design their business models to adapt to these events. He proposes several key rules to "navigate situations in which the unknown predominates and our understanding is limited". Even though he is a former derivatives trader, his terminology is amazingly compatible to ecological terms for how societies adapt to a scarcity of resources (see Box 14). The key rules Taleb [81] mentions expand upon some of the ideas mentioned in our book. We recommend that the reader read Taleb's book to understand his key rules in more detail [81]:

- Rule 1: Think of the economy as being more like a cat than a washing machine
- Rule 2: Favor businesses that benefit from their own mistakes, not those whose mistakes percolate into the system.
- Rule 3: Small is beautiful, but it is also efficient.

- Rule 4: Trial and error beats academic knowledge.
- Rule 5: Decision makers must have skin in the game.

Today, U.S. citizens continue to practice the western European approach to make social, environmental and economic decisions. This should change if sustainable practices are to be the goal. The western world practices continue to mostly be non-naturally based. If the western world continues business as usual, they will be antifragile using Taleb's terminology. In our view, this approach is untenable in a "boom and bust" cycle world. It is not sustainable.

Moving beyond the western European approach requires people to take some "cultural lessons". This is not just about putting the fork on the left side of your plate at the table. This is a behavior that needs to be adopted that does not care if the fork is on the right or left side of a plate. It is one's behavior that is key but also it is important how one links a culture to nature. Culture and traditions can be the base that allows societies to adapt to cycles of environmental and social change.

The traditional people have customarily been on the fringe throughout all of this "progress" and industrialization. They have been principally bypassed and this was their choice too. They also have maintained a foundation that was informed by their traditional knowledge to guide the future decisions that were quite often made. There is often a tension between the traditional factions and the more modern factions in Indian lands. Some tribal systems are broken or in disarray because they dropped their traditional ideas and culture to become "civilized" (see section 2.1.3). These tribes need to re-immerse themselves in their history. Despite all the variations that exist for how a Native American will behave in Indian country, these people still have a lot to teach the western world.

The western world does not need to search far to find these alternative models about how to connect culture with nature and how to balance decisions made for society and the environment. It is important that the western world does not continue "searching for the holy grail", e.g., does not assume that one model is satisfactory for every situation.

This book is about the diversity of native peoples and why the world needs to take a closer look at their many practices. They have adapted to natural boom and bust cycles. These types of cycles are the norm. They also differ at each location so you cannot adopt one model and expect it to work well in each situation and all the time. A one model approach to decision-making will only create *chaos*. Unfortunately, the tendency of the western world citizens is to search for the "one good solution" that will work everywhere.

We end this book with a summary of a research article just available online in the Proceedings of the National Academy of Science cycles of salmon abundance. If we look at salmon—cultural food for Native Americans—and the cycles that they follow, the long-term adaptation of a native people to salmon is pretty clear. Native Americans did not respond to short-term cycles of salmon populations but to cycles that lasted up to 200 years [145]. Western world scientists have revealed [145] the following:

> "Now work led by University of Washington researchers reveals those decadal cycles may overlay even more important, centuries-long conditions, or regimes, that influence fish productivity. Cycles lasting up to 200 years were found while examining 500-year records of salmon abundance in Southwest Alaska. Natural variations in the abundance of spawning salmon are as large those due to human harvest."
>
> ... "We've been able to reconstruct what salmon runs looked like before the start of commercial fishing. But rather than finding a flat baseline—some sort of long-term average run size—we've found that salmon runs fluctuated hugely, even before commercial fishing started. That these strong or weak periods could persist for sometimes hundreds of years means we need to reconsider what we think of as 'normal' for salmon stocks"...
>
> ... "Surprisingly, salmon populations in the same regions do not all show the same changes through time. It is clear that the salmon returning to different rivers march to the beat of a different—slow—drummer", said Daniel Schindler, UW professor of aquatic and fishery sciences and coauthor of the paper (in the Jan. 14 online early edition of the Proceedings of the National Academy of Sciences).
>
> "The implications for management are profound", Schindler said. "While it is convenient to assume that ecosystems have a constant static capacity for producing fish, or any natural resource, our data demonstrate clearly that capacity is anything but stationary. Thus, management must be ready to reduce harvesting when ecosystems become unexpectedly less productive and allow increased harvesting when ecosystems shift to more productive regimes."

This publication supports how Native Americans adapt to longer cycles of salmon runs. They did not over-exploit salmon at a time when these populations were high. This also is what the Aleuts have done so well. Though Aleuts are the top predators [144] in Alaska, they do not over-exploit resources at any given time. They are sensitive to their habitat and recognize 'when' they need to shift to consuming another resource. They are holistically managing their food.

Native people, who faced unexpected challenges when their lands and resources were conquered, give the world a road map for how to live with nature and to make equitable decisions. People who survive after losing their lands and resources are especially worth emulating. We started our book writing about how some native peoples survived several hundred years after being conquered but did not become part of the new melting pots, e.g., Native Americans, Finns and some of the indigenous communities living in present-day Indonesia. These people have important messages to offer the western world. We have many examples of civilizations that did not survive [146]. They collapsed and became part of another melting pot and never even approached becoming "a salad bowl". This is history that we do not want to emulate as described next [146],

> "Historical collapse of ancient states or civilizations has raised new awareness about its possible relevance to current issues of sustainability, in the context of global change... 12 case studies of societies under stress, of which seven suffered severe transformation. Outcomes were complex and unpredictable. Five others overcame breakdown through environmental, political, or socio-cultural resilience, which deserves as much attention as the identification of stressors. Response to environmental crises of the last millennium varied greatly according to place and time but drew from traditional knowledge to evaluate new information or experiment with increasing flexibility, even if modernization or intensification were decentralized and protracted. Longer-term diachronic experience offers insight into how societies have dealt with acute stress, a more instructive perspective for the future than is offered by apocalyptic scenarios."

One take-home message of this book is that society needs to adapt to longer-term cycles in nature. It needs to factor in the impacts of the volatility or "boom and bust" cycles on what are sustainable practices. We do not have the luxury of continuing to practice the former colonialism practices that still emerge as acceptable practices in some parts of the world. It is time that the western world acknowledged the contribution of native peoples to sustainable practices. Native peoples have a robust diversity of models and sustainability practices that can be used to make social, business and environmental decisions. It is also worth looking at the long surviving practices of native people who have not sold themselves for economic gain and for whom nature is sacred.

**Coyote Essentials**

- The context for leadership is learned from ancestors and historians, it is not learned from reading books.
- Manage natural resources recognizing that volatility in resource supplies is the norm and economic decisions need to be made in a "boom and bust" cycle world.

- Traditional knowledge does not provide specific solutions for a specific problem but an approach for humans to follow that specifically includes nature and humanizes the decision process.
- Do not force people to become a melting pot where individual is lost in a generic group identity. Let people live and behave in a salad bowl where everyone's native customs, values and beliefs are included and accepted. Take advantage of the strengths and abilities that diversity yields and do not replicate people to behave in the same manner.
- People who are specialists will have a harder time surviving under the boom-and bust cycles.

## Chapter 11

# Summary of All Book Coyote Essentials

The root of sustainable practices is maintaining one's link to nature through one's culture since culture creates diversity and diversity equals resilience.

- Maintaining one's cultural link to nature is the key to developing and implementing sustainable natural resource management practices.
- Resource scarcity always exists somewhere in the world.
- Colonialism models are typically/routinely used by those who do not own the resources.

**Fig. 11.1** Illustration depicting the Native American sustainability practices portfolio. Illustration credit: Ryan Rosendal.

- Only practicing good or bad behavior is not sustainable.
- Diverse cultural practices are adapted to specific locations.
- Cultures and traditions are not irrelevant even for technologically advanced societies.
- Language, ceremonies and taboos keep cultures alive and linked to nature.
- For humans to take better care of the earth, art and technology need to have equal emphasis.
- Technology by itself will not produce sustainable nature practices but a balance between technology and spirituality will.
- Traditional knowledge does not provide solutions for a specific problem but an approach for humans to follow that includes nature and humanizes the decision process.
- Western world science developed ecosystem and adaptive management 20 years ago that is similar to Native American nature practices developed hundreds of years ago.
- Western world science decision process is better able to make choices when dealing with a scarcity of land resources but not with a scarcity of knowledge. Native American practices deal with scarcity of land resources and scarcity of knowledge equally well.
- Tribal elders are the Google Search Engines of pre-Columbian knowledge.
- The culture and environmental location of each community/tribe determine which sustainability practices they find acceptable/developed/value.
- A person's culture determines their view of what constitutes a home and what may be considered personal property. Many tribes consider wherever they are currently located as home while Europeans tend to consider their home as a specific location or place.
- Native Americans believe that the Creator made the earth as a place for them to live and thrive. They do not view the earth as a resource to be used as a commodity.
- Empty/Vacant undisturbed land has significant ecological, spiritual and cultural values.
- Tribes are willing to adopt most new technologies as long as they do not negatively impact or conflict with their cultural values.
- Europeans do not consider/take into consideration cultural values in determining which technologies are acceptable. Europeans do rely on technology to resolve societal and economic problems.
- Technology is compatible with nature if resources are used efficiently and wisely, and managed lands are not "decoupled" from nature/retain their overall natural components or are restored to natural condition.
- Consider every component in the ecosystem and ecoregion when making land and resource management decisions.
- Recognize the biases you have and how they influence your decisions.

- Just because an old fish is wrapped in decorative paper doesn't mean it won't be smelly and bad to eat.
- Slow down and think for the 7th generation so you do not misdiagnose nature problems.
- Do not ignore science older than the day you were born.
- Really talk to your grandmother and grandfather, and learn from them.
- Economic choices are more than fancy equations or elegant theories.
- The Native American business model forgoes emphasis on maximizing the generation of profits in order to maintain all of their resources.
- Sustainable natural resource management has to be regionally controlled and framed in a local culture and traditions.
- Include all vested stakeholders in the management of natural resources and really collaborate.
- Never marginalize the up-river people that live and manage natural resources on their lands.
- The context for leadership is learned from ancestors and historians, it is not learned from reading books.
- Manage natural resources recognizing that volatility in resource supplies is the norm and economic decisions need to be made in a "boom and bust" cycle world.
- Traditional knowledge does not provide specific solutions for a specific problem but an approach for humans to follow that specifically includes nature and humanizes the decision process.
- Do not force people to become a melting pot where individual is lost in a generic group identity. Let people live and behave in a salad bowl where everyone's native customs, values and beliefs are included and accepted. Take advantage of the strengths and abilities that diversity yields and do not replicate people to behave in the same manner.
- People who are specialists will have a harder time surviving under the boom-and bust cycles.

# References


[1] Vogt KA, Patel-Weynand T, Shelton M, et al. Sustainability Unpacked—Food, Energy and Water for Resilient Environments and Societies. London, UK and Washington, D.C., USA: Earthscan, 2010.

[2] United Nations Environment Programme (UNEP). Global Environment Outlook: GEO4 environment for development. Valletta, Malta: Progress Press Ltd, 2007. (Accessed January 1, 2013, at http://www.unep.org/geo/geo4/report/ GEO-4_Report _Full_en.pdf.)

[3] Tóth SF and McDill ME. Finding Efficient Harvest Schedules under Three Conflicting Objectives. Forest Science 2009; 55:117-131.

[4] International Forum on Globalization (IFG). Indigenous Peoples and Globalization Program. San Francisco, CA, USA: IFG, 2012. (Accessed January 1, 2013, at http://www.ifg.org/programs/indig.htm.)

[5] Vidal J. The great green land grab. UK: The Guardian, 12 February 2008. (Accessed January 1, 2013, at http://www.guardian.co.uk/environment/2008/feb/13/conservation.)

[6] Sunderland T. Going once, going twice... The great green land grab. Bogor, Indonesia: POLEX (blog for Center for International Forestry Research (CIFOR)), 2012. (Accessed January 1, 2013, at http://www.blog.cifor.org/10770/going-once-going-twice-the-great-green-land-grab/#.UOONUm80V8E.)

[7] World Bank. Growth and $CO_2$ emissions: How do different countries fare? Washington, D.C., USA: The World Bank, Environment Department, 2007. (Accessed January 2, 2013, at http://siteresources.worldbank.org/INTCC/214574-1192124923600/21511758/CO2DecompositionfinalOct2007.pdf.)

[8] Miller RJ. Reservation "Capitalism": Economic Development in Indian Country. Santa Barbara, CA, USA: ABC-CLIO, LLC, 2012.

[9] Fowler WR. Inca Empire. Microsoft Encarta. Microsoft® Encarta® Online Encyclopedia, 2000.

[10] Mann CC. 1491: New Revelations of the Americas Before Columbus. New York, NY, USA: Alfred A Knopf, 2005.

[11] Winters RK. The Forest and Man. New York, NY, USA: Vantage Press, 1974.

[12] Maier P, Smith MR, Keyssar A, Kevles DJ. Inventing America: A history of the United States—Volume 1. New York and London: WW Norton & Company, 2003.

[13] de Blij H. Why Geography Matters: Three Challenges Facing America—Climate Change, The Rise of China, and Global Terrorism. Oxford, UK: Oxford University Press, 2005.

[14] Anderson MK. Tending the Wild: Native American Knowledge and the Management of California's Natural Resources. Berkeley, CA, USA: University of California Press, 2006.

[15] Trefts DC. Canadian and American policy making in response to the first multispecies fisheries crisis in the Greater Gulf of Maine region. In: Hornsby SJ, Reid JG, eds. New England and the Maritime Provinces: Connections and Comparisons. Montreal, Quebec, Canada; McGill-Queen's University Press 2004:207-411.

[16] Redford KH. Ecologically Noble Savage. Cambridge, MA, USA: Cultural Survival Quarterly 1991; 15(1). (Accessed January 2, 2013, at http://www.culturalsurvival.org/ourpublications/csq/article/the-ecologically-noble-savage.)
[17] Diamond J. Collapse: How Societies Choose to Fail or Survive. New York, NY, USA: Penguin Group, 2005.
[18] Franklin B. Poor Richard's Almanack. Philadelphia, PA, USA: American Philosophical Society, 1746.
[19] Baker LA, ed. The Water Environment of Cities. Boston, MA, USA: Springer Scientific, 2009.
[20] Reisner M. Cadillac Desert: The American West and Its Disappearing Water. New York, NY, USA: Penguin Books, 1986.
[21] Perla BS. The differential impact of protection on social and ecological resilience of mountain and lowland societies in the Skagit watershed, USA. PhD Dissertation. Seattle, WA, USA: University of Washington, 2008. (Accessed January 1, 2013, at http://books.google.com/books ? id=Xf4mC4SzzKYC&printsec = frontcover#v=onepage&q&f=false.)
[22] Fitriandi A. Apakah Orang Sunda Mengenal Lapar? Indonesia: Blog at WordPress. com, 2007. (Accessed January 1, 2013, at http://indonesiasejahtera.wordpress.com/2007/08/08/apakah-orang-sunda-mengenal-lapar/.)
[23] Secretariat of the Convention on Biological Diversity. Drinking Water, Biodiversity and Development: A Good Practice Guide. Montreal, Canada, 2012. (Accessed January 1, 2013, at http://www.unwater.org/wwd10/downloads/cbd-good-practice-guide-water-en.pdf.)
[24] Gleick PH. Water and Conflict: Fresh Water Resources and International Security. Oakland, CA, USA: International Security 1993; 18 (1):79-112. (Accessed December 30, 2012, at http://www.pacinst.org/reports/international_security_ gleick_1993. pdf.)
[25] Cooley H, Gleick PH. Climate-proofing transboundary water agreements. Hydrol Sci J 2011; 56(4):711-718.
[26] Mehr F. The politics of water. In: Zichichi A, Ragaini RC. eds. International Seminar on Nuclear War and Planetary Emergencies. 30th session. Ettore Majorana International Centre for Scientific Culture, Erice, Italy: World Scientific Publishing Co. Pte. Ltd., 2004:258.
[27] George W. Coyote Finishes the People: A collection of Indian coyote stories, new & old, telling about the evolution of human consciousness. North Charleston, SC, USA: CreateSpace Independent Publishing Platform, 2011.
[28] Hall RL. The Coquille Indians: Yesterday, Today and Tomorrow. Words & Pictures Unlimited Restoration Edition, 1991.
[29] Vogt KA, Honea JM, Vogt DJ, et al. Forests and Society: Sustainability and Life Cycles of Forests in Human Landscapes. Wallingford, UK: CABI, 2006.
[30] Monticello. Jefferson's Quotations on Nature and the Environment. Charlottesville, VA, USA: The Jefferson Monticello website, The Thomas Jefferson Foundation, Inc. 1996-2012. (Accessed December 31, 2012, at http://www.monticello.org/site/jefferson/quotations-nature-and-environment.)
[31] Paullin CO, Wright JK. Atlas of the Historical Geography of the United States. Washington, D.C., USA and New York, NY, USA: Carnegie Institute of Wash-

ington, American Geographical Society of New York, 1932. http://railroads.unl.edu/documents/view_document.php?id=rail.str.0239.

[32] Beach A. The Gaelic Symphony. Piano Concerto, 1896.

[33] Sandweiss MA. American Progress (by Gast J). Amherst, MA, USA: Amherst College, 1872. (Accessed January 1, 2013, at http://picturinghistory.gc.cuny.edu/item.php?item_id=180,accessed 8_2_12.)

[34] Taylor EK. "Some Fruits of Solitude: wise sayings on the conduct of human life" by W Penn. Scottdale, PA, USA: Herald Press, 2003.

[35] Smyth AH. The Writings of Benjamin Franklin. New York, NY, USA: Macmillan, 1905-1907.

[36] Jackson A. Second Annual Message to Congress, 6 December 1830. (Accessed December 30, 2012, at http://millercenter.org/president/speeches/detail/3634.)

[37] Gover K. An Apology. Department of the Interior at the Ceremony Acknowledging the 175th Anniversary of the Establishment of the Bureau of Indian Affairs, 2000. (Accessed December 30, 2012, at http://www.tahtonka.com/apology.html.)

[38] Echo-Hawk WR. In the Courts of the Conqueror: The 10 Worst Indian Law Cases Ever Decided. Golden, CO, USA: Fulcrum Publishing, 2010.

[39] Edwards C. Indian Lands, Indian Subsidies, and the Bureau of Indian Affairs. Washington, D.C., USA: CATO Institute, 2012. (Accessed December 30, 2012, at http://www.downsizinggovernment.org/interior/indian-lands-indian-subsidies.)

[40] US National Park Service (NPS). Yellowstone National Park: History and Culture, 2012. (Accessed January 1, 1013, at http://www.nps.gov/yell/historyculture/index.htm; with additional photos at: http://www.nps.gov / features/yell/slidefile/history/indians/Images/02726.jpg; http://www.nps.gov/features/yell/slidefile/history/1872_1918/military/Images/02802.jpg;http://www.nps.gov /features/yell/slidefile/his tory/1872_1918/peopleevents/Images/02949.jpg.)

[41] Websters Dictionary. Campaign hats. (Accessed January 2, 2013, at http://www.websters-dictionary-online.org/definitions/campaign+hat.)

[42] Ojibwa. National Parks & Native Americans: Yellowstone. Native American Netroots, 2010. (Accessed December 30, 2012, at http://www.nativeamericannet.roots.net/diary/688/national-parks-american-indians-yellowstone.)

[43] Indian Self-Determination and Education Assistance Act (ISDEAA). U.S. Public Law 93-638, 1975. (Accessed December 30, 2012, at http://tm112.community.uaf.edu/files/2010/09/Self-DeterminationAct-19751.pdf.)

[44] Archdale J. A new description of that fertile and pleasant province of Carolina: with a brief account of its discovery, settling, and the government thereof to this time. With several remarkable passages of divine providence during my time. London, England: London (Printed for J. Wyat) 1707; American culture series 103:9. (Accessed December 30, 2012, at http://www.worldcat.org/title/new-description-of-that-fertile-and-pleasant-province-of-carolina-with-a-brief - account - of-its-discovery-settling-and-the-government-thereof-to-this-time-with -several - remarkable-passages-of-divine-providence-during-my-time/oclc/6738053.)

[45] Craven K. Native spirituality today. Seattle Times, Seattle, WA, USA, November 2012; A12 News.

[46] United Nations (UN) Economic and Social Council. Indigenous Peoples and Boarding Schools: A Comparative Study. Permanent Forum on Indigenous Issues, Ninth session, New York, April 19-30, 2010. (Item 3, 4, 7 of the provisional agenda. Special theme: "Indigenous peoples: development with culture and identity: articles 3 and 32 of the United Nations Declaration on the Rights of Indigenous Peoples"); 2010:E/C.19/2010/16. (Accessed January 1, 2013, at http://www.un.org/esa/socdev/unpfii/documents/E.C.19.2010.16%20EN.pdf.)

[47] Adams DW. Education for Extinction—American Indians and the Boarding School Experience 1875-1928. Lawrence, Kansas, USA: University Press of Kansas, 1995.

[48] George LJ. Why the need for the Indian Welfare act? Journal of Multi-Cultural Social Work 1997; 5:165-175.

[49] Svenson S. Bison. Pierre, SD. USA: South Dakota Department of Game, Fish and Parks, Division of Wildlife, 1995. (Accessed January 1, 2013, at http://www3.north ern.edu/natsource/MAMMALS/Bison1.htm.)

[50] World Commission on Dams (WCD). Dams and Development: A New Framework for Decision-making—The Report of the World Commission on Dams. London, UK & Sterling, VA, USA: Earthscan Publications Ltd., 2000. (Accessed January 2, 2013, at http://www.dams.org/report.)

[51] Northwest Power and Conservation Council. Ceremony of Tears. Portland, OR, USA, 2009. (Accessed December 30, 2012, at http://www.nwcouncil.org/history/images/ceremony.jpg.)

[52] Ortolano L, Kao-Cushing K. Grand Coulee Dam 70 Years Later: What Can We Learn? Carfax Publishing (Routledge), International Journal of Water Resources Development, 2002;18(3):373-390.

[53] BIA. Frequently Asked Questions. (Accessed July 28, 2012, at www.bia.gov/FAQs/index.htm.)

[54] Rigdon PH. Indian Forest: Land in Trust (Case 3.4). In: Vogt KA, Honea JM, Vogt DJ, et al., eds. Forests and Society—Sustainability and Life Cycles of Forests in Human Landscapes. Wallingford, OX, UK: CABI International, 2007:105-109.

[55] Spokesman Review. Regional Tribes Included in Federal Settlement. Spokane, WA, USA: Spokesman-Review, 2012. (Accessed January 1, 2013, at http://m.spokesman.com/stories/2012/apr/12/regional-tribes-included-in-federal-settlement/.)

[56] Virtanen L, DuBois T. Finnish Folklore (Studia Fennica Folkloristica—Book 9). Helsinki, Finland: Finnish Literature Society, 2000; in association with the Seattle, WA, USA: University of Washington Press, 2001.

[57] Schiff JW, Moore K. The impact of sweat lodge ceremony on dimensions of well-being. American Indian & Alaska Native Mental Health Research, 2006; 13(3): 48-69.

[58] Moerman DE. Native American Medicinal Plants: An Ethnobotanical Dictionary —The medicinal uses of more than 3000 plants by 218 Native American tribes. Portland, OR, USA: Timber Press, Inc, 2009.

[59] Ngabut CY. Sastra Lisan Suku Kenyah Bakung di Desa Long Apan Baru. In: Eghenter C, Sellato B, eds. Kebudayaan dan Pelestarian Alam: Penelitian Interdisipliner di Pedalaman Kalimantan, Jakarta. Indonesia: Indonesia Forest Protec-

tion and Nature Conservation (PHPA), the Ford Foundation, WWF for Nature Indonesia, 1999:487-504.
[60] McWhorter LV. Yellow Wolf: His Own Story. Caldwell, ID, USA: The Caxton Printers, Ltd, 1940.
[61] Jacques M. When China Rules of the World: The End of the Western World and the Birth of A New Global Order. New York, NY, USA: The Penguin Press, 2009.
[62] Adimihardja K. Kasepuhan Yang Tumbuh Di Atas Yang Luruh: Pengelolaan Lingkungan Secara Tradisional Di Kawasan Gunung Halimun, Jawa Barat. Bandung: Tarsito, 1992.
[63] Sastrawijaja A, Adimihardja K. Sistem Pengetahuan Dan Teknologi Rakyat: Subsistensi Dan Pembangunan Berwawasan Lingkungan Di Kalangan Masya- rakat Sunda Di Jawa Barat. Bandung: Ilham Jaya, 1994. (Indigenous knowledge system of Sundanese people in Jawa Barat Province.)
[64] Berry, JK, Gordon JC. Environmental Leadership: Developing Effective Skills and Styles. Washington D.C., USA: Island Press, 1993.
[65] Vale TR. Fire, Native Peoples, and the Natural Landscape. Washington D.C., USA: Island Press, 2002.
[66] Anderson MK. Traditional Ecological Knowledge: An Important Facet of Natural Resources Conservation. NRCS, United States Department of Agriculture and Natural Resource Conservation Services. (Accessed December 29, 2012, at http://npdc.usda.gov/pdf/0105_tek_report.pdf.)
[67] Siscawati M. Social Movements and Scientific Forestry: Examining the Community Forestry Movement in Indonesia. Seattle, WA, USA: University of Washington Unpublished Dissertation, 2012.
[68] Lu F, Gray C, Bilsborrow RE, et al. Contrasting Colonist and Indigenous Impacts on Amazonian Forests. Conservation Biol. 2010; 24(3):881-885.
[69] Smith LT. Decolonizing Methodologies: Research and Indigenous Peoples. London, UK: Zed Books Ltd., 2005.
[70] Kirk R, Daugherty R. Exploring Washington Archaeology. Seattle, WA, USA: University of Washington Press, 1978.
[71] Bonnicksen T, Anderson K, Lewis H, Kay C, Knudson R. Native American Influences on the Development of Forest Ecosystems. In: Szaro RC, Johnson NC, Sexton WT, Malk AJ, eds. Ecological Stewardship: A common reference for ecosystem management. Vol. 2. Oxford, UK: Elsevier Science Ltd, 1999:439-470.
[72] Kimmerer, R. Native Knowledge for Native Ecosystems. Journal of Forestry 2000; 98(8):4-9.
[73] Gordon J, Franklin J, Morishima G and Johnson KN. The ecological condition of Indian forests: the IFMAT View. Evergreen Winter 2005-2006; 6-10. (Accessed December 30, 2012, at http://evergreenmagazine.com/magazine/article/The_Ecological_Condition_of_Indian_Forests_The_Ifmat_View.html.)
[74] Gordon J. Ecosystem Management: An Idiosyncratic Overview.. In: Aplet GH, Johnson N, Olson JT, Sample VA, eds. Defining Sustainable Forestry. Covelco, CA, USA: Island Press, 1993:240-244.
[75] Dininny S. Lake Roosevelt drawdown approved by state. Seattle, WA, USA: Seattle Times, 2008. (Accessed December 30, 2012, at http://seattletimes.com/html/localnews/2008204557_columbia26.html.)

[76] Morell SF. A review of "Tending the Wild: Native American Knowledge and the Management of California's Natural Resources, by MK Anderson". Washington, D.C., USA: Weston A. Prize Foundation for Wise Traditions in Food, Farming, and the Healing Arts, 2012. (Accessed December 31, 2012, at http://www.westonaprice.org/thumbs-up-reviews/tending-the-wild-by-m-kat-anderson.)
[77] The Economist. Kings of the carnivores. New York, NY, USA: The Economist Group, The Economist online, April 30, 2012. (Accessed January 1, 2013, at http: //www.economist.com/blogs/graphicdetail/2012 / 04 / daily-chart-17? utm_source=Sightline+Newsletters&utm_campaign=9fa6b79667-SightlineDaily&utm_medium=email.)
[78] Sobel D. Look, Don't Touch: The problem with environmental education. Great Barrington, MA, USA: Orion magazine, July/August 2012. (Accessed January 1, 2013, at http://www.orionmagazine.org/index.php/articles/article/6929.)
[79] Kaufman L. 2012. When babies don't fit plan, question for zoos is, now what? New York, NY, USA: The New York Times, August 2, 2012. (Accessed December 30, 2012, at http://www.nytimes.com/2012/08/03/science/zoos-divide-over-contraception-and-euthanasia-for-animals.html?pagewanted=all&_r=0.)
[80] Schumacher EF. Small is Beautiful: Economics as if People Mattered. Harper Perennial, New York, NY, USA: Harper & Row Publishers, Inc., 1973.
[81] Taleb NN. 2012. Learning to love volatility: In a world that constantly throws big, unexpected events our way, we must learn to benefit from disorder. New York, NY, USA: The Wall Street Journal, November 23-25, 2012:14-15. (Accessed January 1, 2013, at http://online.wsj.com/article/SB10001424127887324735104578120953311383448.html.)
[82] Gordon JC, Berry JK. Environmental Leadership Equals Essential Leadership: Redefining Who Leads and How. New Haven, CT, USA: Yale University Press, 2006.
[83] Wilton D. Word Myths: Debunking Linguistic Urban Legends. New York, NY, USA: Oxford University Press, 2004.
[84] Murner T. Narrenbeschwörung (or Appeal to Fools). Hamburg, Germany: Ex Bibliotheca Gymnasii Altonani, 1512. (Accessed December 30, 2012, at http://simple.wikipedia.org/wiki/File:Murner.Nerrenbeschwerung.kind.jpg.)
[85] Richards MP. A brief review of the archaeological evidence for Palaeolithic and Neolithic subsistence. European Journal of Clinical Nutrition 2002; 56(12):1270-1278.
[86] Online Etymology Dictionary. Indian. (Accessed December 30, 2012, at http://www.etymonline.com/index.php?allowed_in_frame=0&search=Indian&searchmode=none.)
[87] Treuer D. Kill the Indians, Then Copy Them. New York, NY, USA: The New York Times September 30, 2012:8. (Accessed January 1, 2013, at http://www.nytimes.com /2012 / 09 / 30 / opinion / sunday/kill-the-indians-then-copy-them.html.)
[88] Seymour M. Little Estate on the Prairie. A review of Peter Pagnamenta's "Prairie Fever: British Aristocrats in the American West, 1830-1890." New York, NY, USA: The New York Times, Sunday Book Review June 29, 2012. (Accessed Jan-

uary 1, 2013, at http://www.nytimes.com/2012/07/01/books/review/prairie-fever-by-peter-pagnamenta.html.)

[89] Yglesias M. Bill Koch Building Private Cowboy Town "For His Huge Collection of Western Memorabilia". New York, NY, USA: The Huffington Post, Huff Post Business, August 21, 2012. (Accessed January 2, 2013, at http://www.huffingtonpost.com/2012/08/21/bill-koch-private-town_n_1818529.html.)

[90] Alter A. Your E-Book Is Reading You. New York, NY, USA: The Wall Street Journal, June 20, 2012:D1-D2. (Accessed December 30, 2012, at http://online.wsj.com/article/SB10001424052702304870304577490950051438304.html.)

[91] de Saint-Exupéry A. The Little Prince. Gallimard, ed., New York, NY, USA: Reynal & Hitchcock, Inc., 1943.

[92] Finn H. How to End the Age of Inattention. Could a Yale program for doctors help everyone pick up on the details? New York, NY, USA: The Wall Street Journal, June 1, 2012. (Accessed December 30, 2012, at http://online.wsj.com/article/SB10001424052702303640104577436323276530002.html.)

[93] Bloch M. The Historian's Craft. New York, NY, USA: Vintage Books, Division of Random House, 1953.

[94] Taylor H. What we are afraid of. New York, NY, USA: The Harris Poll#49, 18 August 1999. (Accessed January 1, 2013, http://www.harrisinteractive.com/vault/Harris-Interactive-Poll-Research-WHAT-WE-ARE-AFRAID-OF-1999-08.pdf.)

[95] Sellato B. Innermost Borneo: Studies in Dayak Cultures. Paris, France: SevenOrients, 2002, & Singapore: Singapore University Press, 2002.

[96] Down to Earth. A portrait of indigenous forest management in Sungai Utik. Down to Earth Bulletin Indonesia 2006; No. 70. (Accessed December 30, 2012, at http://www.downtoearth-indonesia.org/story/portrait-indigenous-forest-management-sungai-utik.)

[97] Marglin SA. Why economists are part of the problem. Washington, D.C., USA: The Chronicle Review, 2009. (Accessed December 30, 2012, at http://chronicle.com/weekly/v55/i25/25b00701.htm.)

[98] Krugman P. How did economists get it so wrong? New York, NY, USA: The New York Times, 2 September, 2009. (Accessed December 30, 2012, at http://www.nytimes.com/2009/09/06/magazine/06Economic-t.html?pagewanted=all&_r=0.)

[99] Jackson RB, Carpenter SR, Dahm CN, et al. Water in a Changing World. Issues in Ecology, Ecological Society of America: Number 9, Spring 2001. (Accessed December 30, 2012, at http://www.esa.org/science_resources/issues/TextIssues/issue9.php.)

[100] Gleick PH, Wolff G, Chalecki EL, Reyes R. The New Economy of Water: The Risks and Benefits of Globalization and Privatization of Fresh Water. Oakland, CA, USA: Pacific Institute for Studies in Development, Environment, and Security, 2002. (Accessed December 30, 2012, at http://www.pacinst.org/reports/new_economy_of_water/new_economy_of_water.pdf.)

[101] Indianz.com. Griles to Be Sentenced for Lying About Abramoff. June 26, 2007. (Accessed December 30, 2012, at http://www.Indianz.com/News/2007/003607.asp.)

[102] Wilkins DE. Native American Politics and the American Political System. Lanham, MD, USA: Rowman & Littlefield Publishing Group, 2007.

[103] Parker LS. Native American Estate: The Struggle Over Indian and Hawaiian Lands. Honolulu, HI, USA: University of Hawaii Press, 1989.

[104] Human Development Index (HDI). HDI Rankings for countries in 2007/2008. UNDP, Human Development Reports. (Accessed December 30, 2012, at http://hdr.undp.org/en/statistics/.)

[105] Icelandic Energy Marketing Unit. Lowest Energy Prices. Reykjavik, Iceland, 1997:15. (Accessed January 2, 2013, at http://blog.pressan.is/larahanna/files/2010/03/Lowest_energy_prices_Iceland.pdf.)

[106] KPMG. Introduction to Business Costs in Iceland: A Comparative study of 87 cities in Europe, North America and Japan. Iceland: KPMG Iceland, 2002. (Accessed December 30, 2012, at http://www.mmkconsulting.com/projects/iceland_summary_2002.pdf.)

[107] Swan J. The Icelandic rift industry versus natural splendor in a "progressive" nation. Great Barrington, MA, USA: Orion Magazine, M.G.H. Gilliam, March/April 2004. (Accessed January 1, 2013, at http://www.savingiceland.org/2004/03/the-icelandic-rift-industry-versus-natural-splendor-in-a-progressive-nation-by-jon-swan/.)

[108] Ogmundardottir H. The Shepherds of pjórsárver: Traditional use and hydropower development in the commons of the Icelandic Highland. PhD Dissertation. Uppsala, Sweden: University of Uppsala, 2011.

[109] McKinsey & Company. Charting a Growth Path for Iceland. Copenhagen, Sweden: McKinsey Scandinavia/McKinsey & Company, 2012. (Accessed December 30, 2012, at http://www.mckinsey.com/locations/Copenhagen/our_work/How_We_Work / ~ / media / Images / Page_Images / Offices /Copenhagen/ ICELAND_Report_2012.ashx.)

[110] Pétursdóttir, SK, Björnsdóttir SH, Ólafsdóttir S, Hreggviðsson GÓ. Líffræðilegur fjölbreytileiki í hverum að peistareykjum og í Gjástykki. Reykjavik, Iceland, 2008: Matís Report 39-08. (Accessed December 30, 2012, at http://www. theistareykir.is / static / files / skyrslur_greinar / 2008 / liffradilegur_fjolbreytileiki_-_heil darskyrsla.pdf.)

[111] Vogt KA, Gordon J, Wargo J, et al. Ecosystems: Balancing Science with Management. New York, NY, USA: Springer-Verlag, 1997.

[112] Global Land Project (GLP). Science Plan and Implementation Strategy. Stockholm, Sweden: IGBP Secretariat 2005; IGBP Report No. 53/IHDP Report No. 19. (Accessed December 30, 2012, at http://digital.library.unt.edu/ark:/67531/metadc12009/m1/1/.)

[113] Vogt KA, Larson BC, Vogt DJ, Gordon JC, Fanzeres A. Forest Certification: Roots, Issues, Challenges and Benefits. Boca Raton, FL, USA: CRC Press, 1999.

[114] Rischard J-F. High Noon: 20 Global Problems, 20 Years To Solve Them. New York, NY, USA: Basic Books (Perseus Books Group), 2002.

[115] Yardley W. Tea party blocks pact to restore a west coast river. New York, NY, USA: The New York Times. July 19, 2012: A16. (Accessed January 2, 2013, at http://www.nytimes.com/2012/07/19/us/two-years-after-pact-to-restore-river-no-changes.html?pagewanted=all&_r=0.)

[116] CBO. How Federal Policies Affect the Allocation of Water. The Congress of the United States, Congressional Budget Office. (Accessed December 30, 2012, at http://www.cbo.gov/sites/default/files/cbofiles/ftpdocs/74xx/doc7471/08-07-waterallocation.pdf.)

[117] Wilkinson C. Blood Struggle: The Rise of Modern Indian Nations. New York, NY, USA: W.W. Norton & Company, 2005.

[118] Rawls JJ, Nash GD, Etulain RW. Chief Red Fox Is Dead: A History of Native Americans Since 1945. Belmont, CA, USA: Wadsworth/Thomson Learning, 2001.

[119] Wilson S. Research Is Ceremony: Indigenous Research Methods. Winnipeg, Canada: Fernwood Publishing, 2008.

[120] Payne SB Jr. The Iroquois League, the Articles of Confederation, and the Constitution. The William and Mary Quarterly, 3rd Series, 1996; 53(3): 605-620. (Accessed January 1, 2013, at http://www.jstor.org/discover/10.2307/2947207?uid=2129&uid=2&uid=70&uid=4&sid=21101607937187.)

[121] Weatherford J. Indian Givers: How the Indians of the Americas Transformed the World. New York, NY, USA: Fawcett Books, 1988.

[122] Pandey A and Rao PV. Impact of Globalization on Culture of Sacred Groves: A Revival of Common, But Decay of the Traditional Institution. In Ninth Biennial Conference of IASCP. Zimbabwe; 2002:17-21.

[123] Vogt KA, Vogt DJ, Patel-Weynand T, et al. Why Forest Derived Biofuels can Mitigate Climate Change: The case for C-based energy production. Renewable Energy, 2008.

[124] Kennett DJ, Breitenbach SFM, Aquino VV, et al. Development and disintegration of Maya political systems in response to climate change. Science 2012; 338:788-791.

[125] Cosens B. Transboundary river governance in the face of uncertainty: resilience theory and the Columbia River Treaty. J Land, Resources & Environ Law 2010; 30: 229-265. (Accessed December 30, 2012, at http://www.epubs.utah.edu/index.php/jlrel/article/viewFile/333/273.)

[126] CRITFC (Columbia River Inter-tribal Fish Commission). Columbia River Treaty Tribes. 2012. (Accessed October 8, 2012, at http://www.critfc.org/text/tribes.html.)

[127] CRITFC (Columbia River Inter-tribal Fish Commission). Columbia Basin Fish Accords Ceremony. 2012. (Accessed November 1, 2012, at http://critfc.org/chp/signing.html.)

[128] Gardner L, Spilde KA. Harnessing Resources, Creating Partnerships: Indian gaming and diversification by the Tulalip Tribes of Washington. The Harvard Project on American Indian Economic Development. Malcolm Wiener Center for Social Policy, John F. Kennedy School of Government, Harvard University, 2004. (Accessed December 30, 2012, at http://hpaied.org/images/resources/publibrary/NIGACaseStudyTulalip.pdf.)

[129] Red-Horse V. Quil Ceda Village: Tribal Gaming's Model City. Casino Enterprise Management 2011. November 1, 2011. (Accessed December 12, 2012, at http://www.casinoenterprisemanagement.com/articles/november-2011/quilceda-village- tribal-gaming%E2%80%99s-model-city.

[130] Borrini-Feyerabend G, Pimbert M, Farvar T, Kothari A, Renard Y. Sharing power: Learning by doing in comanagement of natural resources throughout. London: International Institute for Environment and Development, 2004.

[131] Ingles AW, Musch A, Qwist-Hoffman H. The participatory process for supporting collaborative management of natural resources: An overview. Rome: Food and Agriculture Organization of the United Nations, 1999.

[132] Gray B. Conditions facilitating interorganizational collaboration. Human Relations 1985; 38 (10): 911-936.

[133] Williams EM, Effefson PV. Going into partnership to manage a landscape. Journal of Forestry 1997; 95(5):29-33.

[134] Pinchot Institute for Conservation (Pinchot Institute). Partnership with the USDA Forest Service: Improving opportunities and enhancing relationships. Washington DC: Pinchot, 2001.

[135] Selin S, Chavez D. Developing a collaborative model for environmental planning and management. Environmental Management 1995; 19(2):189-195.

[136] Wondolleck JM, Yaffe SL. Making collaboration work: Lessons from innovation in natural resource management. Washington DC: Island Press, 2000.

[137] Armitage D, Berkes F, Doubleday N. Adaptive Co-Management: Collaboration, Ledarning, and Multi-Level Governance. Vancouver, BC, Canada: University of British Columbia Press, 2007. (Accessed January 2, 2013, at http://www.ubcpress.ca/books/pdf/chapters/2007/adaptivecomanagement.pdf.)

[138] Clark DA, Lee DS, Freeman MMR, Clark SG. Polar bear conservation in Canada: Defining policy problems. Arctic 2008; 61(4): 347-360.

[139] Berkes F, Colding J, Folke C. Rediscovery of traditional ecologic knowledge as adaptive management. Ecological Applications 2000; 10(5): 1251-1262.

[140] Kimmerer R. Weaving traditional ecological knowledge into biological education: A call to action. Bioscience 2002; 52(5): 432-438.

[141] Nadasdy P. The politics of traditional ecological knowledge: Power and the integration of knowledge. Arctic Anthropology 1999; 36(1-2): 1-18.

[142] Vogt KA, Scullion JJ, Nackley LL, Shelton M. Conservation Efforts, Contemporary. In: Levin S, ed. Encyclopedia of Biodiversity. 2nd ed. San Diego, CA, USA: Academic Press, 2012.

[143] Rulli MC, Saviori A, D'Odorico P. Global land and water grabbing. Proceedings of the National Academy of Science Early Edition 2012. (Accessed December 25, 2012; http://www.pnas.org/cgi/doi/10.1073/pnas.1213163110.)

[144] Travis J. No omnivore's dilemma for Alaskan hunter-gatherers. Washington, D.C., USA: American Association for the Advancement of Science, Science Now, February 19, 2012. (Accessed January 1, 2013, at http://news.sciencemag.org/sciencenow/2012/02/no-omnivores-dilemma-for-alaskan.html.)

[145] Hines S. Salmon runs boom, go bust over centuries. UW News, January 14, 2012. (Accessed January 14, 2013, at http://www.washington.edu/news/2013/01/14/salmon-runs-boom-go-bust-over-centuries/?utm_source=rss&utm_medium=rss &utm_campai gn=salmon-runs-boom-go-bust-over-centuries.)

[146] Butzer KW, Endfield GH. Critical perspectives on historical collapse. NPAS 2012; 109: 3628-3631. www.pnas.org/cgi/doi/10.1073/pnas.1114772109.

# Index

**H**



**I**



**J**

**T**

# Description of Ledger Art and The Story behind The Book Cover

Ledger art is important to me as an Indian Artist: I do current and past stories on antique papers from old ledger books. My influences come from my culture and traditions from the Colville confederated tribes, which makes up a group of various tribes that belong to the plateau land basin. I also have an A.A. degree in Museum Studies and a B.A. Degree in Art History. I have worked as a Museum manager and taught ledger art to elementary and high school students. I have a passion for my culture and traditions. My hopes are in viewing my ledger art you can remember the stories in my ledgers, because this type of art is pictorial art: it tells a story through pictures. Ledger art is a Traditional Art that originated as far back as Rock paintings and Pictographs: ancient art of Native American Indians. I do stories from significant events from the past and present. I am a Plateau Style Artist.

Here is the story on my ledger horse:

- This art was not made to be political however the ledger paper "Payroll of the Climax Molybdenum Company" caused the ledger drawing to make more of a statement than intended.
- There used to be a mine here that mined molybdenum and as companies go they close down and move. I was told that my spirit horse had a meaning in guarding our mountains from this type of devastation.
- The spirit horse was intended to protect and had mystical powers because the horse is a metamorphism of a buffalo, eagle into a horse, with powers from its markings of hale spots and lightning with the stars to represent power of the spirit world.
- To see this is a good thing, because in our area we like the horses and we are known to be horse people.

Cheryl A Grunlose
Address-708 Holly St. Coulee Dam, WA 99116
cherylgrunlose@hotmail.com
www.facebook.com/cheryl.grunlose.9